W0232328

Titles in This Series

Titles in This Series

Titles in This Series

Primes Associated to an Ideal

CONTEMPORARY MATHEMATICS

102

Primes Associated to an Ideal

Stephen McAdam

AMERICAN MATHEMATICAL SOCIETY • PROVIDENCE, RHODE ISLAND

1980 *Mathematics Subject Classification* (1985 *Revision*). Primary 13A17.

Library of Congress Cataloging-in-Publication Data

McAdam, Stephen.
    Primes associated to an ideal/Stephen McAdam.
    p. cm. — (Contemporary mathematics, ISSN 0271-4132; v. 102)
    Includes bibliographical references (p.).
    ISBN 0-8218-5108-X (alk. paper)
    1. Noetherian rings. 2. Ideals (Algebra) I. Title. II. Series: Contemporary mathematics
(American Mathematical Society); v. 102.
QA251.3.M3726 1989
512'.4—dc20   89-27624
                                                    CIP

To Martha

(It bears repeating.)

ACKNOWLEDGEMENTS

I deeply thank my esteemed colleagues Jack Ratliff and Dan Katz, who contributed immensely to the material presented herein. After writing an earlier monograph, I swore to never again undertake such a task. A large part of the blame for talking me out of that wise oath (which I again take) falls on the heads of the "Purdue Crew", Bill Heinzer, Craig Huneke, and their spate of recent graduates, in particular, Sam Huckaba and Jugal Verma. Nor does Judy Sally escape blameless. Ray Heitmann was, as always, the person who instantly told me the answers to any questions over which I had struggled for weeks.

Despite the marvels of modern microprocessors, turning a floppy disk into camera ready copy is no mean feat. Dave and Joan Sands of Longhorn Copies more than lived up to their titles of Saint Dave and Joan of Arc with their patient and skilled help.

TABLE OF CONTENTS

PREFACE

This text discusses five closely related sets of prime ideals associated to an ideal I in a Noetherian ring, the persistent, asymptotic, quintasymptotic, essential, and quintessential primes of I. The first two of these were studied in [M2]. Since the appearance of that monograph, the other three sets were developed, and more was learned about the first two sets as well. Matters have reached a state in which the known results are scattered throughout some three dozen papers, making it difficult for an interested person to learn the subject. The aim of this work is to present the most important and interesting of the ideas in an efficient manner, easing the burden of those who wish to learn much of, or simply refer to part of what is known concerning these sets.

The background required is little more than a standard year course in commutative ring theory. Thus, the work should be accessible to many graduate students, and I would be delighted should a dissertation or two be conceived while it is being read. While this work is primarily intended for commutative ring theorists, I would like to think that noncommutative ring theorists and algebraic geometers might also find it of interest.

As the ideas studied in [M2] are closely related to the ideas discussed here, there is some overlap between that work and this one. I adopted the following strategy. If a result from [M2] could be reproved in a way which added insight into the relationship between the older and newer ideas, then I included the new proof here. Otherwise, I simply refer to results in [M2] as needed.

1 BASIC RESULTS

Notation. R will always be a Noetherian ring, and I will be an ideal of R. $R(I) = R[u, It]$ with t an indeterminate and $u = t^{-1}$ will be the Rees ring of R with Respect to I. We will denote the integral closure of I by $\overline{I}$. The integral closure of R is R'. If (R, M) is a local ring, R^* will denote its M-adic completion.

$\overline{Q}^*(I) = \{P \in \text{Spec } R \mid I \subseteq P$ and there is a minimal prime q in $R_P{}^*$ with $P_P{}^*$ minimal over $IR_P{}^* + q\}$ (the quintasymptotic primes of I).

$\overline{A}^*(I) = \{P \cap R \mid P \in \overline{Q}^*(uR(I))\}$ (the asymptotic primes of I).

$Q(I) = \{P \in \text{Spec } R \mid I \subseteq P$ and there is a prime $q \in \text{Ass } R_P{}^*$ with $P_P{}^*$ minimal over $IR_P{}^* + q\}$ (the quintessential primes of I).

$E(I) = \{P \cap R \mid P \in Q(uR(I))\}$ (the essential primes of I).

$A^*(I) = \text{Ass } R/I^n$ for all large n (the persistent primes of I). (See [B] or [M2, Corollary 1.5] to see that $A^*(I)$ is well defined.)

(In certain early papers, Q(I) was called E(I), and E(I) was called U(I). In [M2], the phrase asymptotic primes of I was used indiscriminately to refer to either $A^*(I)$ or $\overline{A}^*(I)$.)

(1.1) Lemma. a) Let A(I) represent any one of $\overline{Q}^*(I)$, $\overline{A}^*(I)$, Q(I), E(I), or $A^*(I)$. Let S be a multiplicatively closed subset of R disjoint from the prime P. Then $P \in A(I)$ if and only if $P_S \in A(I_S)$.

b) Let $q \in$ Ass R, and let P be minimal over $I + q$. Then $P \in Q(I)$. If q is a minimal prime of R, then $P \in \overline{Q}^*(I)$.

c) $P \in Q(I)$ (respectively, $P \in E(I)$) if and only if there is a $q \in$ Ass R with $q \subseteq P$ and $P/q \in Q(I + q/q)$ (respectively, $P/q \in E(I + q/q)$). Also, $P \in \overline{Q}^*(I)$ (respectively, $P \in \overline{A}^*(I)$) if and only if there is a minimal prime q of R with $q \subseteq P$ and $P/q \in \overline{Q}^*(I + q/q)$ (respectively, $P/q \in \overline{A}^*(I + q/q)$).

d) $\overline{Q}^*(I) \subseteq Q(I) \cap \overline{A}^*(I)$ and $\overline{A}^*(I) \cup Q(I) \subseteq E(I) \subseteq A^*(I)$.

e) A prime minimal over I is in $\overline{Q}^*(I) \cap \overline{A}^*(I) \cap Q(I) \cap E(I) \cap A^*(I)$.

Proof. a) This is easy when $A(I) = A^*(I)$. For the other cases, it follows from the definitions and the fact that $R(I)_S = R_S(I_S)$.

b) By (a), we may assume that R is local at P. Passing to the completion, we find that P^* is minimal over $IR^* + qR^*$. Let p be a prime of R^* minimal over qR^*. Then $p \in$ Ass R^* and P^* is minimal over $IR^* + p$. By definition, $p \in Q(I)$. If q is minimal in R, then p is minimal in R^*, and so we get that $P \in \overline{Q}^*(I)$.

c) The proofs of the statements involving $\overline{Q}^*(I)$ and $\overline{A}^*(I)$ are analogous to the proofs for $Q(I)$ and $E(I)$, and so we will only present the latter. In these, it does no harm to assume that R is local at P. Let $P \in Q(I)$. Then there is a $p \in$ Ass R^* with P^* minimal over $IR^* + p$. Let $q = p \cap R$, so that $q \in$ Ass R. The completion $(R/q)^*$ can be identified with R^*/qR^*. Also, $p/qR^* \in$ Ass R^*/qR^*. Now $(P/q)^* = P^*/qR^*$ is minimal over $(IR^* + p)/qR^* = (I + q/q)(R^*/qR^*) + p/qR^*$. This shows that $P/q \in Q(I + q/q)$. The converse is similar.

Now suppose that $P \in E(I)$. If $\mathbf{R} = R(I)$, then there is a $P' \in Q(u\mathbf{R})$ with $P' \cap R = P$. By what we have just proved, there is a $q' \in$ Ass $\mathbf{R}$ with $q' \subseteq P'$ and $P'/q' \in Q(u\mathbf{R} + q'/q')$. Let $q = q' \cap R$. It is not hard to see that $q \in$ Ass R, and $q' = qR[u, t] \cap \mathbf{R}$. (That is, q' has the form

$\cdots + qu^2 + qu + q + (q \cap I)t + (q \cap I^2)t^2 + \cdots.)$ It follows that $\mathbf{R}/q'$ is isomorphic to the Rees ring of R/q with respect to $I + q/q$. We will call this Rees ring $\mathbf{R}/\mathbf{q}$. The image of P'/q' under this isomorphism is in $Q(u(\mathbf{R}/\mathbf{q}))$, and that image intersects R/q at P/q. Thus by definition, $P/q \in E(I + q/q)$. The converse is similar.

d) $\overline{Q}^*(I) \subseteq Q(I)$ and $\overline{A}^*(I) \subseteq E(I)$ are obvious from the definitions. The other inclusions can be proved in a variety of ways, and we will use proofs which fit most naturally into our overall presentation. $\overline{Q}^*(I) \subseteq \overline{A}^*(I)$ will be proved in $(1.6)(1)$. $Q(I) \subseteq E(I)$ will be proved in (1.10). $E(I) \subseteq A^*(I)$ will be proved in in (1.17).

e) Using (d), it is enough to show that a prime minimal over I is in $\overline{Q}^*(I)$. However, this follows easily from (b).

The definitions given above show that $\overline{A}^*(I)$ is to $\overline{Q}^*(I)$ as $E(I)$ is to $Q(I)$, an analogy which is often useful, as in the proof of Lemma 1.1(c). In [M2], another definition of $\overline{A}^*(I)$ is used, one which is also very useful. Our next lemma shows that the two definitions are equivalent.

(1.2) Lemma. $\overline{A}^*(I) = \cup \{\text{Ass} R/\overline{I^n} \mid n \geq 1\} = \text{Ass} R/\overline{I^n}$ for all large n.

Proof. [M2, Proposition 3.9] shows the second equality holds. We will let $\overline{A}^{**}(I)$ denote Ass $R/\overline{I^n}$ for all large n. (That is, $\overline{A}^{**}(I)$ is the $\overline{A}^*(I)$ of [M2].) We want to show $\overline{A}^{**}(I) = \overline{A}^*(I)$ as defined above. By [M2, Proposition 3.18(i) $\Leftrightarrow$ (ii)], $\overline{A}^{**}(I) = \{Q \cap R \mid Q \in \overline{A}^{**}(uR(I))\}$. Also, by definition, $\overline{A}^*(I) = \{Q \cap R \mid Q \in \overline{Q}^*(uR(I))\}$. Thus it will suffice to show that $\overline{A}^{**}(uR(I)) = \overline{Q}^*(uR(I))$. Using $(1.1)(c)$ and [M2, Proposition 3.18 (i) $\Leftrightarrow$ (iv)], we see that we may work modulo a minimal prime of $R(I)$. Changing notation, we need that if $0 \neq x \in T$ with T a Noetherian domain, then $\overline{A}^{**}(xT) = \overline{Q}^*(xT)$. Using [M2, Proposition 3.19 (i) $\Leftrightarrow$ (ii) $\Leftrightarrow$ (iii)], and a harmless localization, we see that for a prime Q of T, $Q \in \overline{A}^{**}(xT)$ if and only if Q_Q^* is minimal over $xT_Q^* + q$ for some minimal prime q of T_Q^*. By the definition, this is the same as saying that $Q \in \overline{Q}^*(xT)$. This completes the proof. (See (3.14) for a generalization of the fact that $\overline{A}^*(xT) = \overline{Q}^*(xT)$.)

Notation. Recall that for k an integer, $P^{(k)} = P^k R_P \cap R$. We analogously define $P^{<k>} = \overline{P^k} R_P \cap R$. Now let I and J be ideals. By $I : <J>$ we will mean the eventual stable value of the increasing chain $(I : J) \subseteq (I : J^2) \subseteq (I : J^3) \subseteq \cdots$.

(1.3) Proposition. Let $I \subseteq P \in$ Spec R. The following are equivalent.

1) $P \in Q(I)$.

2) There is an integer $k \geq 1$ such that $P \in$ Ass R/J for every ideal J satisfying $I \subseteq$ Rad J and $J \subseteq P^{(k)}$.

3) There is an integer $k \geq 1$ such that for all $n \geq 1$, $I^n : <P> \not\subseteq P^{(k)}$.

(1.4) Proposition. Let $I \subseteq P \in$ Spec R. The following are equivalent.

1) $P \in \overline{Q}^*(I)$.

2) There is an integer $k \geq 1$ such that $P \in$ Ass R/J for any ideal J satisfying $I \subseteq$ Rad J and $J \subseteq P^{<k>}$.

3) There is an integer $k \geq 1$ such that for all $n \geq 1$, $I^n : <P> \not\subseteq P^{<k>}$.

4) There is an integer $k \geq 1$ such that for all $n \geq 1$, $\overline{I^n} : <P> \not\subseteq P^{<k>}$.

The next lemma is a key step in proving both (1.3) and (1.4).

(1.5) Lemma. Let (R, M) be a local ring, and let $q \in$ Ass R. Then there is a $0 \neq x \in R$ such that for any ideal J with M minimal over J + q, either $x \in J$ or $M \in$ Ass R/J. If q is a minimal prime of R, then x can be chosen outside of q (and hence ouside the nilradical of R).

Proof. Let $z_1 \cap \ldots \cap z_n = 0$ be a primary decomposition of 0 with z_1 primary to q. Pick $x \in (z_2 \cap \ldots \cap z_n) - z_1$. If q is minimal, we may also take x to be outside of q. Suppose J is an ideal, and M is minimal over J + q. Assume $M \notin \mathrm{Ass}\ R/J$. Let P be any prime in $\mathrm{Ass}\ R/J$. Then $q \not\subseteq P$, so that $z_1 \not\subseteq P$. As $xz_1 = 0$, we see that x is in every P-primary ideal. As this is true of all $P \in \mathrm{Ass}\ R/J$, we have $x \in J$.

Proof of $(1.4)(1) \Rightarrow (2)$ and $(1.3)(1) \Rightarrow (2)$. Suppose $P \in \overline{Q}^*(I)$. It does no harm to localize at P (so that $P^{<k>} = \overline{P^k}$). By definition, there is a minimal prime q in R^* with P^* minimal over $IR^* + q$. Pick $x \in R^* - q$ as in (1.5) applied to (R^*, P^*) and q. By [M2, Lemma 3.11], the intersection of $\overline{P^{*k}}$ over all $k \geq 1$ equals the nilradical of R^*, and so we may pick k large enough that $x \notin \overline{P^{*k}}$. Now let J be an ideal of R with $I \subseteq \mathrm{Rad}\ J$ and $J \subseteq \overline{P^k}$). Then P^* is minimal over $JR^* + q$ and $JR^* \subseteq \overline{P^{*k}}$. By (1.5), $P^* \in \mathrm{Ass}\ R^*/JR^*$, so that $P \in \mathrm{Ass}\ R/J$. This proves $(1.4)(1) \Rightarrow (2)$. The proof of $(1.3)(1) \Rightarrow (2)$ is similar, using the Krull intersection theorem in place of [M2, Lemma 3.11].

(1.6) Remarks. 1) We now give part of the proof of (1.1)(d), showing that $\overline{Q}^*(I) \subseteq \overline{A}^*(I)$. Suppose that $P \in \overline{Q}^*(I)$. The preceding shows that statement (1.4)(2) holds. Let k be as in that statement, and suppose $n \geq k$ is an integer. Then $I \subseteq \mathrm{Rad}\ \overline{I^n}$ and $\overline{I^n} \subseteq P^{<k>}$. Therefore, $P \in \mathrm{Ass}\ R/\overline{I^n}$. As this holds for all $n \geq k$, (1.2) shows that $P \in \overline{A}^*(I)$.

2) An argument similar to that for (1) shows $Q(I) \subseteq A^*(I)$.   (In (1.10) and (1.17) we prove the stronger statement $Q(I) \subseteq E(I) \subseteq A^*(I)$. However, the statement $Q(I) \subseteq A^*(I)$ just noted will be needed in order to reach (1.10) and (1.17).)

(1.7) Lemma. Let I and J be two ideals of R. Let $w_1 \cap ... \cap w_r = I$ be a primary decomposition of I, and let $z_1 \cap ... \cap z_m = 0$ be a primary decomposition of 0. Then

a) $I : <J> = \cap w_i$ over those w_i such that $J \not\subseteq Rad\ w_i$.

b) $\cap (I^n : <J>) = \cap z_i$ over those z_i such that $J \not\subseteq Rad\ (I + z_i)$.

c) $\cap (\overline{I^n} : <J>) = \cap q$ over those minimal primes q such that $J \not\subseteq Rad\ (I + q)$.

Proof. (a) is an easy exercise. For (b), since $\cup Ass\ R/I^n$ over all $n \geq 1$ is finite ([M2, Corollary 1.5]), we may pick $x \in J$ such that x is in a prime contained in $\cup Ass\ R/I^n$ if and only if J is contained in that prime. Now (a) shows that $I^n : <J> = I^n : <xR>$ for all $n \geq 1$. Let S be the multiplicatively closed set $S = \{x, x^2, x^3, ...\}$. It is easily seen that $I^n : <xR> = I^n R_S \cap R$ for all $n \geq 1$. The Krull intersection theorem shows that $\cap (I^n R_S \cap R) = \cap z_i$ over those z_i such that $I + z_i$ is disjoint from S. Therefore, it will suffice to show that this last intersection equals $\cap z_i$ over those z_i with $J \not\subseteq Rad\ (I + z_i)$. That is, we want $x \in Rad\ (I + z_i)$ if and only if $J \subseteq Rad\ (I + z_i)$. Since $x \in J$, one direction is clear. For the other, assume $x \in Rad\ (I + z_i)$ and let Q be any prime minimal over $I + z_i$. We need $J \subseteq Q$. Now Q is minimal over $I + Rad\ z_i$. By (1.1)(b) and (1.6)(2), $Q \in A^*(I)$. Since $x \in Q$, the choice of x shows that $J \subseteq Q$, as desired.

To prove (c), since (1.2) and [M2, Proposition 3.17] show that $\cup Ass\ R/\overline{I^n} = \overline{A}^*(I) \subseteq A^*(I)$, we have $\cup Ass\ R/\overline{I^n} \subseteq \cup Ass\ R/I^n$. Thus, with x and S as before, for all $n \geq 1$, $\overline{I^n} : <J> = \overline{I^n} : <xR> = \overline{I^n}\ R_S \cap R$. It follows from [M2, Lemma (3.11)] that $\cap (\overline{I^n}\ R_S \cap R) = \cap q$ over those minimal primes q such that $I + q$ is disjoint from S. The rest of the proof of (c) is identical to that of (b).

We now complete the proofs of (1.3) and (1.4).

Proof of (1.4). We have already proved (1) $\Rightarrow$ (2). Now suppose (2) holds, and let k be as therein. We will show that (3) holds for this k. For any $n \geq 1$, clearly $I \subseteq \text{Rad}(I^n : <P>)$. Also, (1.7)(a) shows that P is not a prime divisor of $I^n : <P>$. By (2), we must have $I^n : <P> \not\subseteq P^{<k>}$.

Thus, (3) holds. Now (3) $\Rightarrow$ (4) is trivial. Finally, suppose (4) holds, but that (1) fails. We will derive a contradiction. It does no harm to assume that R is local at P. As (1) fails, P^* is not minimal over $IR^* + q$ for any minimal prime q of R^*. By (1.7)(c), $\cap(\overline{I^nR^*} : <P^*>)$ over all $n \geq 1$ equals N, the nilradical of R^*. Thus $\cap((\overline{I^nR^*} : <P^*>)/N)$ is zero in R^*/N. As this ring is complete, Chevally's theorem [N3, 30.1] says that with k as in (4), there is an n with $((\overline{I^nR^*} : <P^*>)/N) \subseteq (P^*/N)^k$.

Thus $\overline{I^nR^*} : <P^*> \subseteq P^{*k} + N \subseteq \overline{P^{*k}}$. Now $\overline{I^n} : <P> \subseteq$

$(\overline{I^nR^*} : <P^*>) \cap R \subseteq \overline{P^{*k}} \cap R = \overline{P^k} = P^{<k>}$, using [M2 Lemma 3.15]. This contradicts (4), and completes the proof of (1.4).

Proof of (1.3). This is almost identical to the proof of (1.4), being simpler in that there is no need to work modulo N.

(1.8) Lemma. Let the Noetherian ring T be a faithfully flat extension of R. Then the Rees ring T(IT) is a faithfully flat extension of the Rees ring R(I).

Proof. Let $I = (a_1, ..., a_n)R$, and let $X_1, ..., X_n$ be indeterminates. Note that $T[u, X_1, ..., X_n]$ is faithfully flat over $R[u, X_1, ..., X_n]$. Let K be the kernel of the natural map from $R[u, X_1, ..., X_n]$ onto $R(I) =$

$R[u, a_1/u, ..., a_n/u]$, and let K' be the kernel of the natural map from $T[u, X_1, ..., X_n]$ onto $T(IT) = T[u, a_1/u, ..., a_n/u]$. Since $T[u]$ is faithfully flat over $R[u]$, it is not hard to see that $K' = KT[u, X_1, ..., X_n]$, and the result follows.

(1.9) Proposition. Let the Noetherian ring T be a faithfully flat extension of R. Let A(I) represent any one of $\overline{Q}^*(I)$, $\overline{A}^*(I)$, Q(I), E(I), or $A^*(I)$.

a) If $Q \in A(IT)$, then $Q \cap R \in A(I)$.

b) If $P \in A(I)$ and if $Q \in \operatorname{Spec} T$ is minimal over PT, then $Q \in A(IT)$.

Proof. When $A(I)$ represents $A^*(I)$, the result follows from standard facts, as in [N3, 18.11].

Suppose $A(I)$ represents $Q(I)$. Let $Q \in Q(IT)$, and let k be as in (1.3)(2) applied to Q and IT. Now let J be any ideal of R with $I \subseteq \operatorname{Rad} J$ and $J \subseteq (Q \cap R)^{(k)}$. Then $IT \subseteq \operatorname{Rad} JT$ and $JT \subseteq Q^{(k)}$. Therefore, $Q \in \operatorname{Ass} T/JT$, so that $Q \cap R \in \operatorname{Ass} R/J$. By (1.3), $Q \cap R \in Q(I)$. This proves (a). For (b), (still taking $A(I) = Q(I)$), let $P \in Q(I)$, and let $Q \in \operatorname{Spec} T$ be minimal over PT. Then $R_P \subseteq T_Q$ satisfies the Theorem of Transition, and so is a faithfully flat extension [N3, 19.1]. Note that $\operatorname{Rad} P_P T_Q = Q_Q$. Since $Q(I)$ localizes well, we may add to our original assumptions that (R, P) and (T, Q) are local, and that $\operatorname{Rad} PT = Q$. By definition of $Q(I)$, there is a $p \in \operatorname{Ass} R^*$ with P^* minimal over $IR^* + p$. Thus in T^*, Q^* is minimal over $IT^* + pT^*$. Let $q \in \operatorname{Spec} T^*$ be minimal over pT^*. Then $q \in \operatorname{Ass} T^*$. Since Q^* is minimal over $IT^* + q$, we have that $Q \in Q(IT)$.

For $A(I) = \overline{Q}^*(I)$, the proof is analogous to the last paragraph, using (1.4), and taking p to be minimal in R.

For $A(I) = E(I)$, let $\mathbf{R}$ and $\mathbf{T}$ be the Rees rings of R with respect to I, and T with respect to IT, respectively. Then $\mathbf{T}$ is faithfully flat over $\mathbf{R}$ by (1.8). Suppose that $Q \in E(IT)$. Then there is a $q \in Q(u\mathbf{T})$ with $q \cap T = Q$. By the second paragraph of this proof, $q \cap \mathbf{R} \in Q(u\mathbf{R})$, so that $Q \cap R = (q \cap \mathbf{R}) \cap R \in E(I)$. This proves (a). For (b), suppose $P \in E(I)$, and $Q \in \operatorname{Spec} T$ is minimal over PT. As was argued in the second paragraph of this proof, we may assume that (R, P) and (T, Q) are local, and $\operatorname{Rad} PT = Q$. Since $P \in E(I)$, there is a $p \in Q(u\mathbf{R})$ with $p \cap R = P$. Let q be a prime of $\mathbf{T}$ minimal over $p\mathbf{T}$. By the second paragraph of this proof, $q \in Q(u\mathbf{T})$. Thus, by definition, we have $q \cap T \in E(IT)$.

However, since $\operatorname{Rad} PT = Q$, we must have $q \cap T = Q$, so $Q \in E(IT)$.

For $A(I) = \overline{A}^*(I)$, the arguments are analogous to the previous paragraph.

Question. Let I be an ideal of R, and let the Noetherian ring T be a faithfully flat extension of R. Let Q be a prime in T with $Q \cap R = P$. Let A(I) represent any of $\overline{Q}^*(I)$, Q(I), $\overline{A}^*(I)$, E(I), or $A^*(I)$. Is it true that $Q \in A(IT)$ if and only if $P \in A(I)$ and $Q \in A(PT)$? (This is in analogy to the well known result $Q \in$ Ass T/IT if and only if $P \in$ Ass R/I and $Q \in$ Ass T/PT.)

(1.10) Remark. We may now prove another part of (1.1)(d), showing that $Q(I) \subseteq E(I)$. First, we claim that if P is minimal over I, then $P \in E(I)$. For this, assume that R is local at P, and in the Rees ring R(I), let p be any prime minimal over uR(I). By (1.1)(b), $p \in Q(uR(I))$. Clearly $p \cap R = P$, and so by definition, $P \in E(I)$, proving the claim. Turning to the main argument, let $P \in Q(I)$. We may assume that R is local at P. By definition, there is a $q \in$ Ass R^* with P^* minimal over $IR^* + q$. Thus P^*/q is minimal over $IR^* + q/q$, so that by the claim, $P^*/q \in E(IR^* + q/q)$. By (1.1)(c), $P^* \in E(IR^*)$, and by (1.9)(a), $P \in E(I)$. (Of course, we could have proved $\overline{Q}^*(I) \subseteq \overline{A}^*(I)$ similarly.)

(1.11) Corollary. Let A(I) represent any one of $\overline{Q}^*(I)$, $\overline{A}^*(I)$, Q(I), E(I), or $A^*(I)$. Let X be an indeterminate. Then $A(IR[X]) = \{PR[X] \mid P \in A(I)\}$.

Proof. (1.9)(b) gives one inclusion. Thus, let $Q \in A(IR[X])$, and let $P = Q \cap R$. Then (1.9)(a) shows that $P \in A(I)$. We need $Q = PR[X]$. By (1.1)(d), $Q \in A^*(IR[X])$. Therefore, for large n, we have that Q is a prime divisor of $I^n R[X]$. It is well known that this implies $Q = PR[X]$.

We wish to look at integral extensions, and need a technical lemma supplied by R. Heitmann.

(1.12) Lemma. Let $R \subseteq T$ be any commutative rings, with R a domain, and let $R^\#$ be a faithfully flat extension of R.

a) Suppose that nonzero elements of R are regular in T. Then regular elements in $R^\#$ are regular in $R^\# \otimes_R T$.

b) Suppose that each minimal prime of T contracts to 0 in R, and that $R^\# \otimes_R T$ has only finitely many minimal primes. Then minimal primes of $R^\# \otimes_R T$ contract to a minimal primes of $R^\#$.

Proof. b) Let Z be a minimal prime of $R^\# \otimes_R T$, and let $z = Z \cap R^\#$. Suppose that (b) is false, and that z properly contains a prime w. Using that $R^\# \otimes_R T$ has only finitely many minimal primes, we easily find $b \in z - w$, and $t \in (R^\# \otimes_R T) - Z$, with $bt = 0$. Fixing b, let t be chosen so that when written $t = \Sigma b_i \otimes t_i$, $i = 1, ..., n$, n is minimal among all possibilities. We claim that $((t_2, ..., t_n)R : t_1 R)_R = 0$. Let $x \in ((t_2, ..., t_n)R : t_1 R)_R$. Then tx can be written in the form $\Sigma c_i \otimes t_i$, $i = 2, ..., n$, and since $b(tx) = 0$, the minimality of n shows us that $tx \in Z$. As $t \notin Z$, $x \in Z \cap R$. However, Z is minimal in $R^\# \otimes_R T$, which is faithfully flat over T. Thus $Z \cap T$ is minimal in T, and so by hypothesis, $Z \cap R = 0$. Thus, $x = 0$, proving our claim. Since $bt = 0$, $bb_1 \in ((t_2, ..., t_n)R^\# : t_1 R^\#)_{R^\#} = (((t_2, ..., t_n)R : t_1 R)_R)R^\# = (0)R^\# = 0$. That is, $bb_1 = 0$. If $t' = \Sigma b_i \otimes t_i$, $i = 2, ..., n$, then $0 = bt = b(b_1 \otimes t_1 + t') = bt'$. The minimality of n shows that $t' \in Z$. However, since we have $bb_1 = 0$, and since $b \notin w$, we also have $b_1 \in w \subseteq z \subseteq Z$, so that $b_1 \otimes t_1 \in Z$. Thus $t = b_1 \otimes t_1 + t' \in Z$, which is a contradiction.

a) If false, let b be a regular element of $R^\#$ and let $0 \ne t \in R^\# \otimes_R T$ with $bt = 0$. Fixing b, we again pick $t = \Sigma b_i \otimes t_i$ to have a minimal number of terms. We again claim $((t_2, ..., t_n)R : t_1 R)_R$ is zero. If x is in this ideal, then tx has an expression shorter than that for t, and as $b(tx) = 0$, we must have $tx = 0$. If $x \ne 0$, the hypothesis shows that x is regular in T, and hence in $R^\# \otimes_R T$, so that $t = 0$. This is a contradiction, showing that $x = 0$. Now as in the proof of part (b), $bt = 0$ implies that $bb_1 = 0$. As b is regular in $R^\#$, $b_1 = 0$. This contradicts that the expression for t was as short as possible.

(1.13) Proposition. Let $R \subseteq T$, where the Noetherian ring T is an integral extension of R.

a) Suppose that T is a finite R-module. If $P \in Q(I)$ (respectively, $P \in E(I)$) then there is a $Q \in Q(IT)$ (respectively, $Q \in E(IT)$ with $Q \cap R = P$. If every $z \in$ Ass T satisfies $z \cap R \in$ Ass R, then both converses hold.

b) Suppose that T is a finite R-module. If $p \in \overline{Q}^*(I)$ then there is a $Q \in \overline{Q}^*(IT)$ with $Q \cap R = P$. If every minimal prime of T contracts to a minimal prime in R, then the converse holds.

c) If $P \in \overline{A}^*(I)$, then there is a $Q \in \overline{A}^*(IT)$ with $Q \cap R = P$. If every minimal prime of T contracts to a minimal prime of R, then the converse holds.

Proof. a) We first treat the case for $Q(I)$. It may be assumed that R is local at P. If $P \in Q(I)$, then there is a $q \in$ Ass R^* with P^* minimal over $IR^* + q$. Letting $T^* = R^* \otimes_R T$, T^* is a finite module over R^*, and so there is a $q' \in$ Ass T^* with $q' \cap R^* = q$. If P' is a maximal ideal if T^* with $q' \subseteq P'$, then it is easy to see that P' is minimal over $IT^* + q'$. By (1.1)(b), $P' \in Q(IT^*)$, and so by (1.9), $P' \cap T \in Q(IT)$. Let $Q = P' \cap T$.

For the converse, assume that every $z \in$ Ass T satisfies $z \cap R \in$ Ass R. Let $Q \in Q(IT)$, and let $P = Q \cap R$. We want that $P \in Q(I)$. By (1.1)(c), there is a $z \in$ Ass T with $z \subseteq Q$ and $Q/z \in Q(IT + z/z)$. Let $w = z \cap R$. By hypothesis, $w \in$ Ass R. Again using (1.1)(c), it will suffice to show that $P/w \in Q(I + w/w)$. However, $P/w = (Q/z) \cap (R/w)$. Thus, we may replace R, P, T, and Q, with R/w, P/w, T/z, and Q/z. That is, to the original asumptions, we may add that R and T are domains (with R local at P), which we now do.

Again letting $T^* = R^* \otimes_R T$, T^* is the completion of the semi-local ring T, and is a finite module over R^*. Now the completion $(T_Q)^*$ of T_Q can be identified with $(T^*)_{QT^*}$. From the definition of

$Q \in Q(IT)$, we see that there is a $z' \in Ass\ T^*$ with QT^* minimal over

$IT^* + z'$. Let $w' = z' \cap R^*$. By the going-up theorem, and the fact that

QT^* is the only maximal prime of T^* which contains z' (since T^*/z' is

a complete semi-local domain, and hence is local), we see that P^* is

minimal over $IR^* + w'$. Now since T is a domain, (1.12)(a) shows that

w' consists of zero divisors in R^*, and so there is a $w'' \in Ass\ R^*$ with

$w' \subseteq w''$. Clearly P^* is minimal over $IR^* + w''$. By (1.1)(b),

$P^* \in Q(IR^*)$, and by (1.9), $P \in Q(I)$.

      Continuing with (a), we consider the case of E(I). Let R(I)
and T(IT) be the Rees ring of R with respect to I and T with respect to IT.

Then T(IT) is a finite module over R(I). If $P \in E(I)$, then there is a

$p \in Q(uR(I))$ with $p \cap R = P$. By what we have already proved, there is a

$q \in Q(uT(IT))$ with $q \cap R(I) = p$. Let $Q = q \cap T$. By definition,

$Q \in E(IT)$, and obviously $Q \cap R = P$.

      For the converse, assume that every prime in Ass T

contracts to a prime in Ass R, and that $Q \in E(IT)$. Since a prime in

Ass R(I) has the form $zR[u, t] \cap R(I)$ for some $z \in Ass\ R$, (and similarly

for T(IT)), we see that primes in Ass T(IT) contract to primes in

Ass R(I). As $Q \in E(IT)$, there is a $q \in Q(uT(IT))$ with $q \cap T = Q$.

By what we already have proved, if $p = q \cap R(I)$, then $p \in QuR(I))$.

Thus $Q \cap R = p \cap R \in E(I)$.

b) The proof of (b) is analogous to the proof given for Q(I) in (a).

c) If we assumed that T was a finite R-module, then we could use (b),
and give a proof analogous to the proof for E(I) in (a). Without the finite
restriction, we use the characterization of $\overline{A}^*(I)$ given in (1.2), and [M2,
Proposition 3.22].

(1.14) Remarks. 1) We do not know if part (b) above really requires that
T be a finite R-module. We suspect that it does not, both in view of part
(c), and in view of the general analogy that Q(I) and E(I) are to finite
module extensions as $\overline{Q}^*(I)$ and $\overline{A}^*(I)$ are to integral extensions. (For
instance, [M2, Chapter X] constitutes a good example of that analogy.)

2) Part (a) does require a finite module. [FR, Proposition 3.3] gives a 2-dimensional local domain (R, M) such that the integral closure R' of R is local, and such that R^* contains a depth 1 prime $q \in$ Ass R. Let $0 \neq b \in R$. As M^* is minimal over $bR^* + q$, $M \in Q(bR)$. If part (a) did not require the finiteness condition, we would have that M lifts to a prime N in $Q(bR')$. By (1.1)(d), $N \in A^*(bR')$. This would imply that N is a prime divisor of $b^n R'$ for some $n \geq 1$. As R' is a Krull domain, we would have that N has height 1. However, as R' is local, N is the only prime lying over M, and so N has height 2, giving a contradiction.

The definitions of $\overline{A}^*(I)$ and $E(I)$ are given via $\mathbf{R} = R(I)$. We now show that $A^*(I)$ can be characterized via $\mathbf{R}$, in a somewhat similar way. This will be useful in completing the proof of (1.1)(d), and later in chapter 12.

(1.15) Proposition. $A^*(I) = \{p \in$ Spec $R \mid I \subseteq p$ and either $p \in$ Ass R or there is a $P \in$ Ass $\mathbf{R}/u\mathbf{R}$ with It $\nsubseteq P$ and $P \cap R = p\}$.

Proof. Consider the sequence Ass R/I, Ass I/I^2, Ass I^2/I^3, $\cdots$. By [M2, Corollary 1.4], this sequence eventually stabilizes to a set denoted $B^*(I)$. Clearly $B^*(I) \subseteq A^*(I)$, and [M2, Proposition 2.2] shows that $A^*(I) - B^*(I) \subseteq$ Ass R. Now [M2, Proposition 1.12] shows that if $I \subseteq p \in$ Ass R, then $p \in A^*(I)$, while [M2, Proposition 1.15] shows that $p \in B^*(I)$ if and only if $p = P \cap R$ for some $P \in$ Ass $\mathbf{R}/u\mathbf{R}$ with It $\nsubseteq P$. Combining these facts gives the present result.

(1.16) Lemma. Let $\mathbf{R} = R(I)$ be the Rees ring of R with respect to I. Let $P \in$ Spec $\mathbf{R}$ with $p = P \cap R$. Suppose that It $\subseteq P$. If $P \in Q(u\mathbf{R})$, then $p \in$ Ass R. If $P \in \overline{Q}^*(u\mathbf{R})$, then p is minimal in R.

Proof. Suppose $P \in Q(u\mathbf{R})$. We may localize at $\mathbf{R} - p$, and assume that R is local at p. Since It $\subseteq P$, P is the unique maximal homogeneous prime of $\mathbf{R}$. We will now show that we may assume that R is complete. Let R^*

be the completion of R. Now if $\mathbf{R}^* = \mathbf{R}^*[u, I\mathbf{R}^*t]$, then $\mathbf{R}^*$ is the Rees ring of $\mathbf{R}^*$ with respect to $I\mathbf{R}^*$, and is a flat extension of $\mathbf{R}$. Let M be the maximal homogeneous prime in $\mathbf{R}^*$. Let $S = \mathbf{R} - P$. Then $\mathbf{R}^*_S$ is flat over $\mathbf{R}_S$, and since M_S lies over the unique maximal ideal P_S of $\mathbf{R}_S$, we see that $\mathbf{R}_S \subseteq \mathbf{R}^*_S$ is a faithfully flat extension. Now $P_S \in Q(u\mathbf{R}_S)$, and so by (1.9), there is an $N \in Q(u\mathbf{R}^*_S)$ with $N \cap \mathbf{R}_S = P_S$. Of course $N \cap \mathbf{R}^* \in Q(u\mathbf{R}^*)$ by (1.1)(a). As $P \subseteq N$, we easily see that $N \cap \mathbf{R}^* = M$. This shows that $M \in Q(u\mathbf{R}^*)$. Now $p = (M \cap \mathbf{R}^*) \cap \mathbf{R}$. In order to achieve our goal of showing that $p \in$ Ass R, it will suffice to show that $M \cap \mathbf{R}^* \in$ Ass $\mathbf{R}^*$. As we have $M \in Q(u\mathbf{R}^*)$ and $I\mathbf{R}^*t \subseteq M$, we have shown that we may assume R is complete.

We return to the notation in the hypothesis, (with the added assumption that R is a complete local ring with maximal ideal p). We next show that we may assume R is a domain. By (1.1)(c), there is a $Q \in$ Ass $\mathbf{R}$, with $Q \subseteq P$ and with $P/Q \in Q(u\mathbf{R} + Q/Q)$. Let $q = Q \cap \mathbf{R}$. Then $q \in$ Ass R, and $Q = q\mathbf{R}[u, t] \cap \mathbf{R}$. Let $\mathbf{R}^\# = (R/q)[u, (I + q/q)t]$, the Rees ring of R/q with respect to $I + q/q$. Then $\mathbf{R}/Q$ is isomorphic to $\mathbf{R}^\#$. Let $P^\#$ be the image of P/Q under this isomorphism. Since $P/Q \in Q(u\mathbf{R} + Q/Q)$, we have $P^\# \in Q(u\mathbf{R}^\#)$. Since $It \subseteq P$, we see that $(I + q/q)t \subseteq P^\#$. Also, $P^\# \cap R/q = p/q$. In order to achieve our goal that $p \in$ Ass R, it will suffice to show that p/q is zero in R/q, since $q \in$ Ass R. Thus we have shown that we may assume that R is a domain.

We return to our original notation with the added assumption that R is a complete local domain with maximal ideal p. Our goal is now to show that $p = 0$. Being a complete local domain, R is analytically unramified. Thus $\mathbf{R}'$, the integral closure of $\mathbf{R}$, is a finite R-module. By (1.13)(a), there is a $P' \in Q(u\mathbf{R}')$ with $P' \cap \mathbf{R} = P$ (so $P' \cap R = p$). By (1.6)(2) $P' \in Q(u\mathbf{R}') \subseteq A^*(u\mathbf{R}') =$ Ass $\mathbf{R}'/u\mathbf{R}'$, and as $\mathbf{R}'$ is a Krull domain, height $P' = 1$. Now the complete local domain R satisfies the altitude formula. Thus height $p + t_1 =$ height $P' + t_2$, where t_1 is the transcendence degree of $\mathbf{R}'$ over R (which we see equals 1, since $\mathbf{R}'$ is integral over $\mathbf{R}$, and $\mathbf{R} \subseteq R[u, t]$), and t_2 is the transcendence degree of $\mathbf{R}'/P'$ over R/p (which we see to be 0, since $\mathbf{R}'/P'$ is integral over

R/P, and $\mathbf{R}/P = R/p$ since It $\subseteq$ P). Therefore, height p + 1 =

height $P' + 0 = 1$. This shows that p = 0 as desired, proving the first
conclusion of the lemma.

Now suppose that $P \in \overline{Q}^*(u\mathbf{R})$. By (1.1)(c), there is a

minimal prime $Q \subseteq P$ with $P/Q \in \overline{Q}^*(u\mathbf{R} + Q/Q)$. Let $q = Q \cap R$. Then q

is minimal in R, and $Q = qR[u, t] \cap \mathbf{R}$. Let $\mathbf{R}^\#$ be the Rees ring of R/q
with respect to I + q/q. We see that $\mathbf{R}/Q$ is isomorphic to $\mathbf{R}^\#$, and if $P^\#$ is

the image of P/Q, then $(I + q/q)t \subseteq P^\# \in \overline{Q}^*(u\mathbf{R}^\#)$, and $P^\# \cap R/q = p/q$.

We already know that $\overline{Q}^*(u\mathbf{R}^\#) \subseteq Q(u\mathbf{R}^\#)$. Therefore, the first
conclusion of this result, which we have just proved, shows that p/q must
be zero in the domain R/q. Thus p = q is minimal in R.

(1.17) Remark. We complete the proof of (1.1)(d) by showing that

$E(I) \subseteq A^*(I)$. Suppose $p \in E(I)$. Then there is a $P \in Q(u\mathbf{R})$, $\mathbf{R} = R(I)$,

with $P \cap R = p$. Now by (1.6)(2), $P \in A^*(u\mathbf{R}) = Ass \ \mathbf{R}/u\mathbf{R}$. If It $\not\subseteq$ P, then

by (1.15), we have $p \in A^*(I)$. If It $\subseteq$ P, then (1.16) shows that $p \in Ass \ R$,

so again (1.15) shows $p \in A^*(I)$.

(1.18) Proposition. Let R be a domain, and let $0 \neq I = (a_1, ...a_n)R$. Let p

be a prime containing I. Then $p \in E(I)$ (respectively, $p \in \overline{A}^*(I)$), if and

only if for some a_i, there is a $q \in E(a_iA)$ (respectively, $q \in \overline{A}^*(a_iA)$)

with $q \cap R = p$, where $A = R[a_1/a_i, ..., a_n/a_i]$.

Proof. We will give the argument for E(I), the other being similar. Let
$\mathbf{R} = R(I)$. Suppose that $p \in E(I)$. Then there is a $P \in Q(u\mathbf{R})$ with

$P \cap R = p$. As $p \neq 0$, (1.16) shows that It $\not\subseteq$ P. Thus for some i, $a_it \notin$ P.

As $a_j/a_i = a_jt/a_it \in \mathbf{R}_P$, we see that $A \subseteq \mathbf{R}_P$. Since $a_i \in I \subseteq p \subseteq P$, $a_i\mathbf{R}_P$
is a proper ideal of $\mathbf{R}_P$. Therefore $IA = a_iA$ is a proper ideal of A.
Let $\mathbf{A} = A[u, a_it]$ be the Rees ring of A with respect to a_iA. We easily see

that $\mathbf{R} \subseteq \mathbf{A} \subseteq \mathbf{R}_P$. If $Q = P \cap \mathbf{A}$, then $\mathbf{A}_Q = \mathbf{R}_P$. Now since $P \in Q(u\mathbf{R})$,

$P_P \in Q(u\mathbf{R}_P)$. Thus $Q_Q \in Q(u\mathbf{A}_Q)$, so that $Q \in Q(u\mathbf{A})$. Let $q = Q \cap A$.
By definition, $q \in E(a_i A)$, and obviously $q \cap R = p$.

Conversly, suppose that $q \in E(a_i A)$ and $q \cap R = p$. There is a
$Q \in Q(u\mathbf{A})$ with $Q \cap A = q$. (1.16) shows that $a_i t \notin Q$. Let $Q \cap \mathbf{R} = P$.
Then $a_i t \in \mathbf{R} - P$, and so $\mathbf{R} \subseteq \mathbf{A} \subseteq \mathbf{R}_P$. We also see that $\mathbf{A}_Q = \mathbf{R}_P$. The
rest of the argument is as before.

(1.19) Remark. The preceding result is adapted from work of
K. Whittington. An analogous result does not hold for $Q(I)$ (or for
$\overline{Q}^*(I)$). In (3.14) we see that if a is a nonzero divisor in R then (since a is
an essential sequence) $Q(aR) = E(aR)$. In particular, with notation as
in (1.18), $Q(a_i A) = E(a_i A)$. If a result analogous to (1.18) held for $Q(I)$,
we would get $Q(I) = E(I)$, which is false.

(1.20) Proposition. Suppose for all maximal ideals M of R, the
completion $R_M{}^*$ has no embedded prime divisors of zero. Then for all
ideals I of R, $E(I) = \overline{A}^*(I)$.

Proof. We already have $\overline{A}^*(I) \subseteq E(I)$. Let $p \in E(I)$. We will show that
$p \in \overline{A}^*(I)$. Let M be a maximal ideal containing p. We may assume
that R is local at M. By (1.9), we may assume that R is complete. By two
uses of (1.1)(c) (and the hypothesis that primes in Ass R are minimal),
we may assume that R is a complete local domain. Let $I = (a_1, ..., a_n)R$.
Let a_i, A, and q be as in (1.18), so that $q \in E(a_i A)$ and $q \cap R = p$. As A is
finitely generated over R and R is analytically unramified, the
integral closure A' is a finite A-module. By (1.13)(a), q lifts to a prime
Q in $E(a_i A')$. By (1.17), $Q \in A^*(a_i A')$. As A' is a Krull domain, height
$Q = 1$. Thus $Q \in \overline{A}^*(a_i A')$. (1.13)(c) shows $q \in \overline{A}^*(a_i A)$, and (1.18)
shows $p \in \overline{A}^*(I)$.

(1.21) Corollary.   The following are equivalent.

i) $P \in E(I)$ (respectively, $P \in \overline{A}^*(I)$).

ii) $P \in E(IJ)$ for any regular ideal J (respectively, $P \in \overline{A}^*(IJ)$ for any ideal J with height $J > 0$).

iii) $P \in E(cI)$ for any $c \in R$ with c not in any prime in Ass R (respectively, $P \in \overline{A}^*(cI)$ for any $c \in R$ with c not in any minimal prime of R).

Proof. The statements involving $\overline{A}^*(I)$ are proved equivalent in [M2, Proposition 3.26]. As for the statements involving E(I), (ii) $\Rightarrow$ (iii) is obvious For (i) $\Rightarrow$ (ii) and (iii) $\Rightarrow$ (i), it does no harm to assume that R is local at P. (For instance, no problem arises if $J_P = R_P$). Also, we may assume that R is complete, by (1.9). By (1.1)(c), we may work modulo some prime in Ass R, and so we may assume that R is a complete local domain. Now (1.20) shows that in such a ring, $E(I) = \overline{A}^*(I)$, and so (i) $\Rightarrow$ (ii) and (iii) $\Rightarrow$ (i) for E(I) follow from the analogous statements for $\overline{A}^*(I)$.

If R is a local ring, let a(I) be the analytic spread of I. The next result is proved in [M2, Proposition 4.1].

(1.22) Proposition. Let $I \subseteq P \in$ Spec R. If height $P = a(I_P)$, then $P \in \overline{A}^*(I)$. If R_P is quasi-unmixed, then the converse holds.

Question. Is there any result similar to (1.22) holding for E(I)?

Most of this chapter comes from [KR1], [M2], [M5], and [MR1]. (The reader may also find Sharp's paper [Sp] of interest.)

2 EXAMPLES

Our first three examples will all deal with the same ring. Let K be a field, and X and Y be indeterminates. Let (R', N_1, N_2) be K[X, Y] localized at the complement of $(X, Y) \cup (X, Y+1)$, with $N_1 = (X, Y)R'$ and $N_2 = (X, Y+1)R'$. Let $R = K + (N_1 \cap N_2)$. Then R is a local ring, its maximal ideal is $M = N_1 \cap N_2$, its integral closure is R', and R' is a finite R-module, since $X \in M$ and $XR' \subseteq R$. Also let $q = YR' \cap R$ and $p = XR' \cap R$. Note that $p = XR' \neq XR$.

(2.1) Example. Claim: $A^*(XR) = \{p, M\}$, and $E(XR) = Q(XR) = \overline{A}^*(XR) = \overline{Q}^*(XR) = \{p\}$. Since p is minimal over XR, it is in each of our sets. Since the only primes containing XR are p and M, by (1.1)(d) it will suffice to show that $M \in A^*(I) - E(I)$. As $YX \in M$ but $Y \notin R$, we have that $YX \in R - XR$. As $MY \subseteq M$, we have $M(YX) \subseteq XR$. Thus $M = (XR : YXR) \in \text{Ass } R/XR = A^*(XR)$, the last equality coming from the fact that X is a regular element of R. It remains to show that $M \notin E(XR)$. If $M \in E(XR)$, then by (1.13)(a), M would have to lift to a prime in $E(XR')$. However, only N_1 and N_2 in R' lie over M, and so one of these would be in $E(R')$. By (1.1)(d), $E(XR') \subseteq A^*(XR')$, and since R' is a Krull domain, primes in this last set have height 1. This is a contradiction, since N_1 and N_2 have height 2.

(2.2) Example. Claim: $A^*(p) = E(p) = Q(p) = \overline{A}^*(p) = \overline{Q}^*(p) = \{p\}$. Obviously p is in each of these sets. As M is the only other prime containing p, by (1.1)(d) it will suffice to show that $M \notin A^*(p)$. If

$M \in A^*(p)$, then for large n we have $M \in$ Ass R/p^n. However, $p = XR'$, so that $p^n = X^n R' = X^n R' \cap R$. Since $M \in$ Ass $R/(X^n R' \cap R)$, M lifts to a prime divisor of $X^n R'$ in R'. As R' is a Krull domain, this says that M lifts to a height 1 prime in R'. This is false, completing the argument. (We just saw that $M \notin$ Ass R/p^n for all $n \geq 1$. Thus p^n is p-primary. Also, since $\overline{A}^*(p) = \{p\}$, (1.2) shows $\overline{p^n}$ is p-primary for $n \geq 1$.)

(2.3) **Example.** Claim: $A^*(q) = E(q) = Q(q) = \overline{A}^*(q) = \overline{Q}^*(q) = \{q, M\}$. Since q is obviously in each of these sets, and since M is the only other prime containing q, by (1.1)(d) it will suffice to show that $M \in \overline{Q}^*(q)$. Since $q \subseteq M$, $qR' \subseteq N_2$. Pick a prime q' of R' with $qR' \subseteq q' \subseteq N_2$, and q' minimal over qR'. Suppose that $q' \neq N_2$. Then $q' \cap R \neq M$, and so we must have $q' \cap R = q$. Now $Xq' \subseteq q' \cap R = q \subseteq YR'$. As X is not in the prime ideal YR', we have $q' \subseteq YR'$. Now both q' and YR' lie over q, so by incomparability, we have $q' = YR'$. Thus, $Y \in q' \subseteq N_2$. However, $Y + 1 \in N_2$, giving a contradiction. This shows that $q' = N_2$, so that N_2 is minimal over qR'. By (1.1)(e), $N_2 \in \overline{Q}^*(qR')$, and by (1.13)(b), $M \in \overline{Q}^*(q)$. This completes the argument. (Since $M \in A^*(q) \cap \overline{A}^*(q)$, M is a prime divisor of q^n and $\overline{q^n}$ for all large n. Since $\overline{Q}^*(q) \neq \{q\}$, (4.3) (with S=R-q) shows that for some $k \geq 1$, $q^{(m)} \not\subseteq \overline{q^k}$ for all $m \geq 1$.)

(2.4) **Example.** We construct a case in which $\overline{Q}^*(I) \neq \overline{A}^*(I)$, and $Q(I) \neq E(I)$. Let $R = K[X, Y]_{(X, Y)}$. Let $M = (X, Y)R$ and $I = XM$. Since $M \in \overline{A}^*(M)$, by (1.21) $M \in \overline{A}^*(XM)$. (Alternately, since the analytic spread of M is 2, it is easily seen by the Hilbert polynomial characterization of analytic spread that the analytic spread of XM is 2. Thus (1.22) shows $M \in \overline{A}^*(XM)$.) We now show that $M \notin \overline{Q}^*(XM)$. Suppose that $M \in \overline{Q}^*(XM)$. Then for some minimal prime q in the completion R^*, we have M^* minimal over $XMR^* + q$. Obviously, M^* is minimal over $XR^* + q$. Therefore, depth $q = 1$. However, it is well known that this R has no depth 1 minimal prime in its completion.

Therefore $\overline{A}^*(XM) \neq \overline{Q}^*(XM)$. Since $M \notin \overline{Q}^*(XM)$, and since R^* has no embedded prime divisors of zero, the definitions show $M \notin Q(XM)$. Also, (1.20) shows $M \in \overline{A}^*(XM) = E(XM)$. Thus $E(XM) \neq Q(XM)$.

(2.5) Example. We construct a case in which $\overline{Q}^*(I) \neq Q(I)$. Let $R = K[X, Y, Z]/X(X, Y)$, let x, y, and z be the images of X, Y, and Z, and let $M = (x, y, z)R$, $P = (x, y)R$, $Q = xR$, and $I = zR$. Since P is a prime divisor of zero, and M is minimal over $zR + P$, $M \in Q(zR)$ by (1.1)(b). Suppose that $M \in \overline{Q}^*(zR)$. By (1.1)(c) and the fact that Q is the only minimal prime of R, we see that $M/Q \in \overline{Q}^*(zR + Q/Q)$. Since R/Q is just K[Y, Z], this says that $(Y, Z)K[Y, Z] \in \overline{Q}^*(ZK[Y, Z])$. By (1.1)(d), $(Y, Z)K[Y, Z] \in A^*(ZK[Y, Z])$. As K[Y, Z] is a Krull domain, we get that $(Y, Z)K[Y, Z]$ has height 1, which is patently false. Thus $M \in Q(zR) - \overline{Q}^*(zR)$.

(2.6) Example. Here is an example more subtle than the previous one, in that the interesting prime divisor of zero is hidden in the completion. We refer again to the 2-dimensional local domain R mentioned in (1.14)(2). The completion R^* has a depth 1 prime divisor of zero, but each of its minimal primes has depth 2. ([M2, Proposition 3.19] is of interest here.) If $0 \neq b \in R$, then we easily see that $M \in Q(bR) - \overline{Q}^*(bR)$. (Remark: [BR] shows how to construct a local ring R such that Ass R^* has a predetermined amount of pathology.)

(2.7) Example. Let $R = K[X, Y]$, with K a field, and X and Y indeterminates. Let $I = (X^2Y, XY^2)R$. We claim $\overline{Q}^*(I) = Q(I) = \{XR, YR\}$, and $\overline{A}^*(I) = E(I) = A^*(I) = \{XR, YR, (X, Y)R\}$. Since XR and YR are minimal over I, they are in $\overline{Q}^*(I) \subseteq Q(I)$. If $P \in Q(I)$, then by definition, P_{P^*} is minimal over $IR_{P^*} + q$ for some $q \in$ Ass R_{P^*}. Either $I \subseteq XR \subseteq P$, or $I \subseteq YR \subseteq P$. Thus P_{P^*} is minimal over either $XR_{P^*} + q$ or $YR_{P^*} + q$. In either case, height $P_{P^*}/q = 1$. However, R is locally unmixed, so that height P_{P^*}/q must equal height P. Therefore, height $P = 1$, and either $P = XR$ or $P = YR$. This shows that $\overline{Q}^*(I) = Q(I) = \{XR, YR\}$, proving the first part of the claim. For the second part, obviously XR and YR are in $\overline{A}^*(I) \subseteq E(I) \subseteq A^*(I)$. Since obviously

$(X, Y)R \in \overline{A}^*((X, Y)R)$, and since $I = XY(X, Y)R$, (1.21) shows that $(X, Y)R \in \overline{A}^*(I)$. Thus we have $\{XR, YR, (X, Y)R\} \subseteq \overline{A}^*(I) \subseteq E(I) \subseteq A^*(I)$. To finish, it will suffice to show that if $Q \in A^*(I)$, then

$Q \in \{XR, YR, (X, Y)R\}$. If height $Q = 1$, then clearly Q is either XR or YR. Thus suppose that height $Q = 2$. We will show that $Q = (X, Y)R$. If not, then in R_Q, either X or Y is a unit. Thus, I_Q is either YR_Q or

XR_Q. As $Q_Q \in A^*(I_Q)$, and I_Q is principal in the Krull domain R_Q, height $Q = 1$, a contradiction. This shows that $\overline{A}^*(I) = E(I) = A^*(I) = \{XR, YR, (X, Y)R\}$.

3 ESSENTIAL AND ASYMPTOTIC SEQUENCES

Definitions. The elements $x_1, ..., x_n$ of R will be called an essential (respectively, asymptotic) sequence if $(x_1, ..., x_n)R \neq R$, and for all $i = 1, ..., n$, x_i is not in any essential (respectively, asymptotic) prime of $(x_1, ..., x_{i-1})R$. If I is an ideal and if $x_1, ..., x_n$ are elements in I which form an essential (respectively, asymptotic) sequence, and if there is no x_{n+1} in I such that $x_1, ..., x_n, x_{n+1}$ is an essential (respectively, asymptotic) sequence, then $x_1, ..., x_n$ will be called a maximal essential (respectively, asymptotic) sequence in I, and n will be called the essential (respectively, asymptotic) grade of I. We will denote this by egr I (respectively, agr I). (That egr I and agr I are well defined is shown in (3.4).) By gr I, we will mean the length of a maximal regular sequence in I, and we will refer to this as the classical grade of I.

(3.1) Lemma. Let $x_1, ..., x_n$ be elements of R with $(x_1, ..., x_n)R \neq R$.

a) x_1 is an essential (respectively, asymptotic) sequence if and only if x_1 is not in any prime divisor of zero (respectively, minimal prime).

b) If $x_1, ..., x_n$ is an essential or asymptotic sequence, then height $(x_1, ..., x_i)R = i$ for all $1 \leq i \leq n$.

c) A regular sequence is an essential sequence, and an essential sequence is an asymptotic sequence. Also, gr I $\leq$ egr I $\leq$ agr I $\leq$ height I.

d) If $x_1, ..., x_n$ is an essential (respectively, asymptotic) sequence in R, and if S is a multiplicatively closed subset of R disjoint from $(x_1, ..., x_n)R$, then $x_1, ..., x_n$ is an essential (respectively, asymptotic) sequence in R_S.

e) If R is local, then $x_1, ..., x_n$ is an essential (respectively, asymptotic) sequence in R if and only if their image in R/q is an essential (respectively, asymptotic) sequence for all q $\in$ Ass R (respectively, minimal primes q of R).

f) If the Noetherian ring T is a faithfully flat extension of R, then $x_1, ..., x_n$ is an essential (respectively, asymptotic) sequence in R if and only if it is an essential (respectively, asymptotic) sequence in T. If I is an ideal in R, then egr I = egr IT (respectively, agr I = agr IT).

g) Let the Noetherian ring T be a finite module (respectively, integral) extension of R. If $x_1, ..., x_n$ is an essential (respectively, asymptotic) sequence in T, then it is an essential (respectively, asymptotic) sequence in R. If also every prime divisor of zero (respectively, minimal prime) of T contracts to a prime divisor of zero (respectively, minimal prime) of R, and if $x_1, ..., x_n$ is an essential (respectively, asymptotic) sequence in R, then it is an essential (respectively, asymptotic) sequence in T. In this case, if I is an ideal of R, then egr I = egr IT (respectively, agr I = agr IT).

Proof. Using (1.1)(d) and the definition of $Q^*(0R)$, we see that $E(0R) \subseteq A^*(0R) =$ Ass $R \subseteq Q(0R) \subseteq E(0R)$. Thus $E(0R) =$ Ass R. Using (1.2), we see that $\overline{A}^*(0R)$ is the set of minimal primes in R. Part (a) follows from these facts. Part (b) follows easily from (1.1)(e). For part (c), since for any ideal I, $\overline{A}^*(I) \subseteq E(I)$, we see that an essential sequence is automatically an asymptotic sequence. Since $E(I) \subseteq A^*(I)$, and since $A^*(I) =$ Ass R/I if I is generated by a regular sequence [Kp, exercise 13, P. 103], it is easily seen that a regular sequence is an essential sequence. For the second part of (c), the first two inequalities follow from the first part of (c), and the third inequality is from part (b). Parts (d) and (e) are easy using (1.1)(a),(c). Part (f) follows from (1.9). Part (g) follows from (1.13)(a),(c).

Recalling that R' is the integral closure of R, it is not hard to produce examples of I an ideal in R with gr I > gr IR' (here assuming R' is Noetherian). Therefore, our next result appears to require the use of essential grade or asymptotic grade.

(3.2) Proposition. Let I be an ideal in R. Then gr I $\leq$ height IR'.

Proof. Let Q be a prime of R' with $IR' \subseteq Q$ and height Q = height IR'.

Let b $\in$ Q with b not in any of the other (finitely many) primes of R'

lying over Q $\cap$ R. Since Q is the only prime of R' lying over Q $\cap$ R[b],

we see that height IR[b] = height IR'. Since R[b] is Noetherian, by

(3.1)(c),(g), gr I $\leq$ egr I = egr IR[b] $\leq$ height IR[b] = height IR'.

Recall that a local ring R is unmixed (respectively, quasi-unmixed) if in the completion R^*, every prime divisor of zero (respectively, minimal prime) has depth equal to Dim R. One goal in this chapter is to illustrate the analogy that Cohen-Macaulay rings are to regular sequences as locally unmixed (respectively, locally quasi-unmixed) rings are to essential (respectively, asymptotic) sequences. One version of the unmixedness theorem for Cohen-Macaulay rings says that if $x_1, ..., x_n$ is a regular sequence in a Cohen-Macaulay ring, and if P is a prime divisor of $(x_1, ..., x_n)R$, then height P = n. We now state an analogous result, the crux of this chapter.

(3.3) Lemma. Let R be locally unmixed (respectively, locally quasi-unmixed). Let $x_1, ..., x_n$ be an essential (respectively, asymptotic) sequence. Let P be an essential (respectively, asymptotic) prime of $(x_1, ..., x_n)R$. Then height P = n and P is minimal over $(x_1, ..., x_n)R$.

Proof. We treat the case that R is locally unmixed, $x_1, ..., x_n$ is an essential sequence, and P is an essential prime of $(x_1, ..., x_n)R$. It does no harm to localize at P, and so we may assume that R is an unmixed local ring. Now by (3.1)(b), height $(x_1, ..., x_n)R = n$. Obviously height P $\geq$ n, and we need only show that height P = n, which implies that P is minimal over $(x_1, ..., x_n)R$. By (1.1)(c), there is a prime p $\in$ Ass R with

p $\subseteq$ P and P/p $\in$ E$((x_1 + p, ..., x_n + p)(R/p))$. By (3.1)(e), $x_1 + p, ..., x_n + p$ is an essential sequence in R/p. As R is unmixed, it is easily seen that height P = height P/p (= height P^*/p^* where $p^* \in$ Ass R^* with $p^* \cap R = p$). Thus we may assume R is a domain.

Let **R** be the Rees ring of R with respect to $(x_1, ..., x_n)R$.

By definition, there is a prime Q of **R** with Q $\in$ Q(u**R**) and Q $\cap$ R = P. Again by definition, in the completion of **R**$_Q$, Q**R**$_Q{}^*$ is minimal over

u**R**$_Q{}^*$ + q, for some q $\in$ Ass **R**$_Q{}^*$. Thus, height Q**R**$_Q{}^*$/q = 1. Since R is unmixed, **R** is locally unmixed (see [N1]), and so we see that

height Q = 1. We consider R ⊆ R[u] ⊆ **R**. As u ∈ Q, clearly

Q ∩ R[u] = (P, u)R[u]. As R is unmixed, it is quasi-unmixed, and so
satisfies the altitude formula, [R2]. Since the transcendence degree of **R**
over R is 1, and since height Q = 1, we see that height P equals the
transcendence degree of **R**/Q over R/P. However, this equals the
transcendence degree of **R**/Q over R[u]/(P,u)R[u] (since the
transcendence degree of R[u]/(P,u)R[u] over R/P is zero). Since **R** is
obtained from R[u] by adjoining the n elements $x_1 t$, ..., $x_n t$, clearly the
transcendence degree of **R**/Q over R/(P,u)R[u] is at most n.
Thus, height P ≤ n. We already have height P ≥ n and so are done in
this case. The other case is argued analogously, using that if R is a
quasi-unmixed local domain, then **R** is locally quasi-unmixed, [R1].

(3.4) Theorem. The following numbers are all equal.

a) The length of any maximal essential (respectively, asymptotic)
sequence in I.

b) min{height $(IR_P^* + q)/q$ | P is a prime of R with I ⊆ P and q is a
prime divisor of zero (respectively, minimal prime) in R_P^*}.

c) min{depth q | q is a prime divisor of zero (respectively, minimal
prime) in R_P^* for some prime P of R with I ⊆ P}.

In particular, the essential (respectively, asymptotic) grade of I is well
defined.

Proof. We will do the essential case, the asymptotic case being argued
analogously. Let x_1, ..., x_n be a maximal essential sequence in I. Let P
be any prime containing I and let q be any prime divisor of zero in the
completion R_P^*. By (3.1)(d), (f), (e), (b), $x_1 + q$, ..., $x_n + q$ is an
essential sequence in R_P^*/q, and generates an ideal of height n. As
that ideal is contained in $(IR_P^* + q)/q$, height $(IR_P^* + q)/q ≥ n$. This
shows that the number expressed in (b) is equal to or greater than n.
Now clearly the number expressed in (c) is equal to or greater than the
number expressed in (b). Therefore, it only remains to show that n is
equal to or greater then the number expressed in (c). Since x_1, ..., x_n is
a maximal essential sequence in I, there is an essential prime Q of

$(x_1, ..., x_n)R$ with $I \subseteq Q$. By (1.1)(a) and (1.9)(b), in the completion R_Q^*, Q_Q^* is an essential prime of $(x_1, ..., x_n)R_Q^*$. By (1.1)(c), there is a prime divisor of zero q in R_Q^* with $q \subseteq Q_Q^*$ and Q_Q^*/q an essential prime of $(x_1 + q, ..., x_n + q)(R_Q^*/q)$. Also, by (3.1)(d), (f), (e), $x_1 + q, ..., x_n + q$ is an essential sequence in R_Q^*/q. Since R_Q^*/q is an unmixed local ring, (3.3) shows that depth $q = n$. Thus n is equal to or greater than the number expressed in (c), and we are done.

In a local ring (R,M), (3.4)(b) can be sharpened, for the minimum it mentions arises when P = M, as we now show.

(3.5) Corollary. Let I be an ideal in a local ring R. Then egr I (respectively, agr I) = min{ height $(IR^* + q)/q$ | q is a prime divisor of zero (respectively, a minimal prime) in R^*}.

Proof. We do the essential case, the asymptotic case being analogous. By (3.4)(b), (letting P = M), we see that egr I is equal to or less then the minimum mentioned in this corollary. For the reverse inequality, let $x_1, ..., x_n$ be a maximal essential sequence in I. Then there is a $P \in E((x_1, ..., x_n)R)$ with $I \subseteq P$. By (1.9)(b) and (1.1)(c), there is a prime Q in R^* with $Q \cap R = P$, and a prime $q \in Ass R^*$ with $q \subseteq Q$, such that Q/q is an essential prime of $(x_1 + q, ..., x_n + q)(R^*/q)$. Since $x_1 + q, ..., x_n + q$ is an essential sequence in the unmixed local ring R^*/q, (3.3) shows that height Q/q = n. As $IR^* + q \subseteq Q$, height $(IR^* + q)/q \le n$. Thus the minimum mentioned in this corollary is equal to or less then egr I.

If $x_1, ..., x_n$ is a regular or essential or asymptotic sequence, by (3.1)(b),(c) height $(x_1, ..., x_i)R = i$ for all $1 \le i \le n$. Concerning converses, it is easily seen that in a Cohen-Macaulay ring, $x_1, ..., x_n$ is a regular sequence if and only if height $(x_1, ..., x_i)R = i$ for $1 \le i \le n$. We now give the analogous results results for essential and asymptotic sequences.

(3.6) Lemma. Let R be locally unmixed (respectively, locally quasi-unmixed). Assume that $(x_1, ..., x_n)R \ne R$. Then $x_1, ..., x_n$ is an essential (respectively, asymptotic) sequence if and only if height $(x_1, ..., x_i)R = i$ for all $1 \le i \le n$. If R is also local, then $x_1, ..., x_n$ is an essential (respectively, asymptotic) sequence if and only if height $(x_1, ..., x_n)R = n$.

Proof. Assume that R is locally unmixed and height $(x_1, ..., x_i)R = i$ for $1 \le i \le n$. We want to show that $x_1, ..., x_n$ is an essential sequence, (the other case being argued analogously). Since in a locally unmixed ring, prime divisors of zero are minimal and since height $x_1 R = 1$, (3.1)(a) shows that x_1 is an essential sequence. By induction, assume that $x_1, ..., x_{i-1}$ is an essential sequence for some $1 < i \le n$. It will suffice to show that x_i is not in any essential prime of $(x_1, ..., x_{i-1})R$. However, this is immediate from (3.3) and the fact that height $(x_1, ..., x_i)R = i$. For the final part of the result, we note that if height $(x_1, ..., x_n) = n$, and if R is a catenary local ring (for instance an unmixed or quasi-unmixed or Cohen-Macaulay local ring), then it is not hard to show that height $(x_1, ..., x_i)R = i$ for $1 \le i \le n$.

We further illustrate the analogy discussed in the paragraph prior to (3.3). The next result is the analogue of a famous result about Cohen-Macaulay rings.

(3.7) Proposition. The following are equivalent.

i) R is locally unmixed (respectively, locally quasi-unmixed).

ii) egr I (respectively, agr I) = height I for all ideals I of R.

iii) egr M (respectively, agr M) = height M for all maximal ideals M of R.

Proof. We will treat the case for locally unmixed rings and essential grade. The case for locally quasi-unmixed rings and asymptotic grade is analogous.

(i) $\Rightarrow$ (ii). Let height I = n. We can find elements $x_1, ..., x_n$ in I with height $(x_1, ..., x_i)R = i$ for all $1 \le i \le n$. Since we have a locally unmixed ring, (3.6) shows $x_1, ..., x_n$ is an essential sequence. Using this and (3.1)(c) $n \le$ egr I $\le$ height I = n, so (ii) holds.

(ii) $\Rightarrow$ (iii) This is trivial.

(iii) $\Rightarrow$ (i). Suppose (iii) holds. To show that R is locally unmixed, it suffices to show that R_M is unmixed for every maximal ideal M. Let height M = n, and by (iii), let x_1, ..., x_n be an essential sequence in M. Let q be a prime divisor of zero in the completion R_M^* of R_M. We need depth q = n. By (3.1)(d), (f), (e), (b), in R_M^*/q, $x_1 + q$, ..., $x_n + q$ generate an ideal of height n. Thus n $\leq$ depth q $\leq$ height M = n.

In (3.10), we give an application of (3.7). The following concept will be useful here, and in chapters 7 and 14.

Definition. Let Q be a prime of R, and let U be an infinite set of primes of R each of which properly contains Q. We will call the pair (Q, U) a conforming pair if any infinite subset U' of U satisfies $\cap\{P \in U'\} = Q$. (Note that if (Q, U) is a conforming pair, and V is an infinite subset of U, then (Q, V) is a conforming pair.)

(3.8) Proposition. Let (Q, U) be a conforming pair in R, and suppose that egr $P_P \leq n$ for all P but finitely many P $\in$ U. Then egr $Q_Q \leq n - 1$. Furthermore, an analogous result holds for asymptotic grade and classical grade.

Proof. We may delete from U those finitely many P for which egr $P_P > n$, and still have a confroming pair. Thus, we suppose that egr $P_P \leq n$ for all P $\in$ U. Suppose the conclusion fails, and let x_1, ..., x_n be elements in R which upon localizing at Q become an essential sequence in Q_Q. Let W = $\cup\{q \in$ E($(x_1, ..., x_{i-1})$R) | $x_i \in$ q}, the union over i = 1, ..., n. Since x_1, ..., x_n is an essential sequence in Q_Q, no q $\in$ W can be contained in Q. Thus, since (Q, U) is a conforming pair, if q $\in$ W then q can be contained in at most finitely many of the primes P $\in$ U. Let us delete from U any prime P which contains a prime q $\in$ W. As W is finite, we have deleted only finitely many P from U, and so still have a conforming pair. Furthermore, we may now assume that if P $\in$ U and q $\in$ W, then q is not contained in P. Thus we easily see that x_1, ..., x_n forms an essential sequence in P_P for all P $\in$ U. Since

egr Pp $\leq$ n, we must have that this is a maximal essential sequence in
Pp. As Pp is maximal in Rp, we must have that

Pp $\in$ E((x_1, ..., x_n)Rp). Thus P $\in$ E((x_1, ..., x_n)R). This holds for all
P $\in$ U. However, U is infinite and E((x_1, ..., x_n)R) is finite, giving a
contradiction. Clearly analogous arguments work for asymptotic and
classical grade.

(3.9) Remark. (3.8) expresses a property central to the very nature of
grade functions. We will define what we mean by an arbitrary grade
function f in chapter 14. We will show that for any grade function f on
the ring R, (3.8) holds for f. In fact, more is true. A function f from the
set of all ideals I_S in all localizations R_S of R to the nonnegative
integers will be shown to be a grade function if and only if f satisfies

(i) f(I_S) = min {f(Pp) | $I_S \subseteq P_S \subseteq$ Spec R_S}, (ii) f(Pp) $\leq$ height P for all
P $\in$ Spec R and (iii) (3.8) holds for f.

(3.10) Proposition. Let Q be a prime of R and let W = {P $\in$ Spec R |
Q $\subseteq$ P and height P/Q = 1}. Suppose that R_Q is unmixed (respectively,
quasi-unmixed) (respectively, Cohen-Macaulay). Then the same is
true of Rp for all but finitely many P $\in$ W.

Proof. Let R_Q be unmixed, and suppose that infinitely many P $\in$ W
have Rp not unmixed. Let U consist of that infinite set of primes P.

Obviously (Q, U) is a conforming pair, (since primes P in U $\subseteq$ W have
height P/Q = 1). By (3.7), we have egr Q_Q = height Q, and

egr Pp < height P for all P $\in$ U (since egr I $\leq$ height I always). By
[M1, Theorem 1], we also have that height P = height Q + 1 for all but

finitely many P $\in$ U. Thus, with finitely many exceptions, for P $\in$ U
we have egr Pp < height P = height Q + 1 = egr Q_Q + 1, so that
egr Pp $\leq$ egr Q_Q. By (3.8), egr Q_Q $\leq$ egr Q_Q - 1. Careful examination of
this last inequality reveals a contradiction. The argument works
equally well for the quasi-unmixed case (using asymptotic grade) and
the Cohen-Macaulay case (using classical grade and the famous
analogue of (3.7)). (In fact, it works for any grade function as
discussed in (3.9). In the abstract setting, it says that if f(Q_Q) =
height Q, then for all but finitely many P $\in$ W, f(Pp) = height P.)

As another application of (3.7), we refer the reader to [Kz2], where D. Katz shows (among other things) that if I is an ideal in a quasi-unmixed local ring, then the associated graded ring of I is locally quasi-unmixed.

(3.11) Remark. Let (R, M) be a local ring, and let $x_1, ..., x_n$ be elements in M. We make two statements whose proofs we leave to the reader.

i) $x_1, ..., x_n$ is an essential (respectively, asymptotic) sequence if and only if height $(x_1 + q, ..., x_n + q)(R^*/q) = n$ for every prime divisor of zero (respectively, minimal prime) in R^*.

ii) If $x_1, ..., x_n$ is an essential (respectively, asymptotic) sequences, then so is every permutation of $x_1, ..., x_n$.

(3.12) Proposition. Let I be an ideal of R, and let b be an element in the Jacobson radical of R. Then egr $I \leq$ egr $(I, b) \leq$ egr $I + 1$, and agr $I \leq$ agr $(I, b) \leq$ agr $I + 1$. (Note: it is well known that gr $I \leq$ gr $(I, b) \leq$ gr $I + 1$.)

Proof. We do the essential case, the asymptotic case being analogous. The first inequality is obvious. If R is local, then the second inequality follows from (3.5), since for any $q \in$ Ass R^*, R^*/q is catenary, and in a catenary local domain, adding one more generator to an ideal can increase the height by at most one. In the general case, let $x_1, ..., x_n$ be a maximal essential sequence from I, and let $P \in E((x_1, ..., x_n)R)$ with I $\subseteq P$. Since b is in the Jacobson radical, there is a maximal ideal M containing (P, b). Since $P \subseteq M$, $x_1, ..., x_n$ is a maximal essential sequence from I_M, so egr $I_M = n$. By the local case, egr $(I, b)_M \leq n + 1$. By (3.1)(d), egr $(I, b) \leq n + 1 =$ egr $I + 1$.

(3.13) Proposition. Let I be generated by n elements. Then egr $I \leq$ agr $I \leq n$. If egr I (respectively, agr I) = n, then I can be generated by an essential (respectively, asymptotic) sequence, (necessarily of length n).

Proof. Since height $I \leq n$, (3.1)(c) gives the first part. The rest of the argument is analogous to the proof of [Kp, Theorem 125] (which does the case for classical grade).

(3.14) Proposition. Let $x_1, ..., x_n$ be an essential (respectively, asymptotic) sequence. Then $E((x_1, ..., x_n)) = Q((x_1, ..., x_n))$ (Respectively, $\overline{A}^*((x_1, ..., x_n)) = \overline{Q}^*((x_1, ..., x_n)))$.

Proof. As always, we do the essential case. As (1.1)(d) gives one inclusion, suppose that $P \in E((x_1, ..., x_n))$. Clearly egr $P_P = n$. By (3.5), there is a $q \in$ Ass R_P^* with height $P_P^*/q = n$. We also know that $x_1 + q, ..., x_n + q$ is an essential sequence in R_P^*/q, and so generates an ideal of height n. P_P^*/q must be minimal over that ideal. Thus, P_P^* is minimal over $(x_1, ..., x_n)R_P^* + q$. By (1.1)(b), $P_P^* \in Q((x_1, ..., x_n)R_P^*)$. By (1.9)(a) and (1.1)(a), $P \in Q((x_1, ..., x_n))$.

(3.15) Remark. Suppose we define a quintessential sequence to be $x_1, ..., x_n$ with $(x_1, ..., x_n)R \neq R$ and x_i not in any quintessential prime of $(x_1, ..., x_{i-1})R$ for $1 \leq i \leq n$. It follows easily from (3.14) that quintessential sequences are identical to essential sequences. Similarly, if we define quintasymptotic sequences in the obvious way, they are identical to asymptotic sequences. [Kp, exercise 13, p. 103] shows that if I is generated by a regular sequence, then $A^*(I) = $ Ass R/I. It follows if we define persistent sequences in the obvious way, they are identical to regular sequences.

(3.16) Remark: If I is an ideal of R, and if $x_1, ..., x_n$ are elements of R with $(I, x_1, ..., x_n)R \neq R$ and with x_i not in any essential prime of $(I, x_1, ..., x_{i-1})R$ for $1 \leq i \leq n$, then $x_1, ..., x_n$ is called an essential sequence over I. Similarly, one can define quintessential, asymptotic, and quintasymptotic sequences over I. (When I = 0, the results are just essential or asymptotic sequences.) Rees first did this for asymptotic sequences over I, [Re], and the idea was further developed by Katz, Ratliff, and the author. We choose not to pursue the details here. The case of asymptotic sequences over I is given in detail in [M2, Chapter VI], while [KR1] shows that analogous results hold for essential sequnces over I. It should be noted that while essential sequences and quintessential sequences are identical (by (3.15)), essential sequences over an arbitrary ideal I are not identical to quintessential sequences over I. The latter are studied in [R3]. (Essential sequences over I are better behaved than quintessential sequences over I.) Similar statements hold for asymptotic and quintasymptotic sequences over I.

Asymptotic sequences were developed soon after the development of asymptotic primes, and were fully treated in [M2, chapter V]. (At that time, essential and quintessential primes had not yet been developed.) It was then wondered if a type of sequence could be developed which stood in relation to prime divisors of zero as asymptotic sequences stood to minimal primes. In order to discuss such a new sequence, the author and Ratliff developed quintessential primes [MR1] (using some ideas of Schenzel [S]), while Katz independently developed essential primes [Kz1]. As (3.15) points out, both paths lead to essential sequences. (However, this author is of the opinion that essential primes are more interesting than quintessential primes.)

4 SCHENZEL'S THEOREMS

Theorems (4.1), (4.2), (4.3), and (4.4), were first proved by P. Schenzel, in [Sc, (3.2), (6.5), (3.5), and (5.6)]. We have made slight improvements to the statements of the theorems, in order to stress their similarities. Our proofs are quite different than the original ones, and are presented so as to further stress their similarities. Finally, our notation is markedly different than Schenzel's.

Notation. Throughout this chapter, S will be a multiplicatively closed subset of R. For J an ideal of R, we use $JR_S \cap R$ to denote $\{x \in R \mid$ there is an $s \in S$ with $sx \in J\}$. (Thus, $JR_S \cap R$ makes sense even when R is not embedded in R_S.)

(4.1) Theorem. The following are equivalent.

a) $S \subseteq R - \cup \{P \in Q(I)\}$.

b) For all $k \geq 1$, there is an $m \geq 1$ with $I^m R_S \cap R \subseteq I^k$.

(4.2) Theorem. The following are equivalent.

a) $S \subseteq R - \cup \{P \in E(I)\}$.

b) There is an $h \geq 0$ such that for all $k \geq 1$, $I^{k+h} R_S \cap R \subseteq I^k$.

(4.3) Theorem. The following are equivalent.

a) $S \subseteq R - \cup \{P \in \overline{Q}^*(I)\}$.

b) For all $k \geq 1$, there is an $m \geq 1$ with $I^m R_S \cap R \subseteq \overline{I^k}$.

b') For all $k \geq 1$, there is an $m \geq 1$, with $\overline{I^m} R_S \cap R \subseteq \overline{I^k}$.

(4.4) Theorem. The following are equivalent.

a) $S \subseteq R - \cup \{P \in \overline{A}^*(I)\}$.

b) There is an $h \geq 0$ such that for all $k \geq 1$, $I^{k+h} R_S \cap R \subseteq \overline{I^k}$.

b') There is an $h \geq 0$ such that for all $k \geq 1$, $\overline{I^{k+h}} R_S \cap R \subseteq \overline{I^k}$.

c) For all $k \geq 1$, $I^k R_S \cap R \subseteq \overline{I^k}$.

c') For all $k \geq 1$, $\overline{I^k} R_S \cap R = \overline{I^k}$.

Proof of (4.1). We will show that (a) and (b) are both equivalent to

c) For all $I \subseteq Q \in \operatorname{Spec} R$ and $k \geq 1$, there is an $m \geq 1$ with
$I^m R_S \cap R \subseteq Q^{(k)}$. (Here, we may restrict to $Q \in Q(I)$.)

(b) $\Leftrightarrow$ (c) One direction is trivial, since if $I \subseteq Q$, then $I^k \subseteq Q^{(k)}$. Thus,
suppose that (c) holds. Consider a primary decomposition $z_1 \cap ... \cap z_n$
of I^k, with z_i primary to Q_i. For sufficiently large k_i, we have
$Q_i^{(k_i)} \subseteq z_i$. By (c), for sufficiently large m_i, we have
$(I^{m_i}) R_S \cap R \subseteq Q_i^{(k_i)} \subseteq z_i$. Letting m be the maximum of $m_1, ..., m_n$,
we see that $I^m R_S \cap R \subseteq z_1 \cap ... \cap z_n = I^k$.

(c) $\Rightarrow$ (a) Suppose that (c) holds, and let $P \in Q(I)$. We need $P \cap S = \emptyset$.

By (1.3)(3), there is a $k \geq 1$ such that for all $m \geq 1$, $I^m : <P> \not\subseteq P^{(k)}$.

By (c), there is an $m \geq 1$ with $I^m R_S \cap R \subseteq P^{(k)}$. For this m, we must

have $I^m : <P> \not\subseteq I^m R_S \cap R$. Let x be in the first set, but not the second.
Then x sends a large power of P into I^m, but nothing in S sends x into
I^m. Thus, nothing in S can be in P, which says (a) holds. (This
argument also proves the parenthetical statement in (c).)

(a) $\Rightarrow$ (c) Suppose (a) holds, and let Q and k be as in (c). If S' is the

image of S in R_Q, then by (1.1)(a), we see that (a) holds for S' and I_Q.

If we can find an m with $I_Q^m (R_Q)_{S'} \cap R_Q \subseteq Q_Q^k$, then intersecting

with R will give $I^m R_S \cap R \subseteq Q^{(k)}$ as desired. Therefore we may
assume that R is local at Q (and that $Q^{(k)} = Q^k$). Now (1.9)(a) implies

that (a) holds for S and IR^*. Since $I^m R_S \cap R \subseteq (IR^*)^m R^*_S \cap R^*$, and

since $Q^{*k} \cap R = Q^k$, we may assume R is complete. Let $z_1 \cap \ldots \cap z_n = 0$
be a primary decomposition of zero. For each i, let p_i be a prime

minimal over $I + \mathrm{Rad}\, z_i$. By (1.1)(b), $p_i \in Q(I)$. By (a), S is disjoint
from p_i, and hence from both $I + z_i$ and $\mathrm{Rad}\, z_i$. Now in R_S,

$(z_1)_S \cap \ldots \cap (z_n)_S = 0$ is a primary decomposition of zero, and since
$I_S + (z_i)_S \neq R_S$, the Krull intersection theorem says for $m \geq 1$,

$\cap I^m R_S = 0$. Since S is disjoint from $\mathrm{Rad}\, z_i$, $1 \leq i \leq n$, S consists of

regular elements in R, and so $\cap(I^m R_S \cap R) = 0$. c) now follows from
Chevally's result, [N3, 30.1].

Proof of (4.2). Let $R(I)$ be the Rees ring of R with respect to I. We see that
for $k \geq 1$, $u^{k+h}R(I)_S \cap R(I)$ has the form $\Sigma J_i t^i$, i an integer, with

$J_i = (I^{k+h+i}R_S \cap R) \cap I^i$ (negative powers of I being R). If (b) holds,

then $J_i \subseteq I^{k+i}$. Thus, $u^{k+h}R(I)_S \cap R(I) \subseteq \Sigma I^{k+i}t^i = u^k R(I)$. Therefore,
(4.1)(b) holds when applied to S and the ideal $uR(I)$, (since for each k, the
m required by (4.1)(b) is $m = k + h$). Therefore (4.1)(a) holds for S and
$uR(I)$, and so S is disjoint from every prime in $Q(uR(I))$. Contracting to
R, we see that S is disjoint from every prime in $E(I)$, so that (a) holds.

Conversly, suppose that (a) holds. Then S is disjoint from every prime in $E(I)$, and so S is disjoint from every prime in $Q(uR(I))$. Thus (4.1)(a) holds when applied to S and the ideal $uR(I)$. Therefore, we may apply (4.1)(b) to S, $uR(I)$, and $k = 1$. We see there is some $m \geq 1$ with

$$u^m R(I)_S \cap R(I) \subseteq uR(I).$$ Now using that u is regular in $R(I)$, an easy

induction shows that for all $k \geq 1$, $u^{m+k-1} R(I)_S \cap R(I) \subseteq u^k R(I)$.

Intersecting with R, we see that $I^{m+k-1} R_S \cap R \subseteq I^k$. Therefore, (b) holds with $h = m - 1$.

We wish to make the proof of (4.3) as similar as possible to that of (4.1). To do this, we need the next lemma.

(4.5) Lemma. If J is an integrally closed ideal in a Noetherian ring R, then J has a primary decomposition in which each component is integrally closed.

Proof. In the integral closure $R(J)'$ of the Rees ring $R(J)$, let $p_1, ..., p_n$ be the primes minimal over $uR(J)'$. If p is one of these primes, then $uR(J)'_p \cap R(J)'$ is an integrally closed p-primary ideal, and $uR(J)'$ is the intersection of all such $uR(J)'_p \cap R(J)'$. Now $(uR(J)'_p \cap R(J)') \cap R$ is an integrally closed $(p \cap R)$-primary ideal, and the intersection of all of these is $uR(J)' \cap R = \overline{J} = J$. Deleting unnecessary components leaves the desired primary decomposition of J.

Proof of (4.3). We will show that (a), (b), and (b′) are all equivalent to both (c) and (c′).

c) For all $I \subseteq Q \in \operatorname{Spec} R$ and $k \geq 1$, there is an $m \geq 1$ with $I^m R_S \cap R \subseteq Q^{<k>}$. (Here, we may restrict to $Q \in \overline{Q}^*(I)$.)

c′) For all $I \subseteq Q \in \operatorname{Spec} R$ and $k \geq 1$, there is an $m \geq 1$ with $\overline{I^m} R_S \cap R \subseteq Q^{<k>}$. (Here, we may restrict to $Q \in \overline{Q}^*(I)$.)

(b) $\Leftrightarrow$ (c) and (b$'$) $\Leftrightarrow$ (c$'$) both use (4.5) and are similar to the proof of

(4.1)(b) $\Leftrightarrow$ (c), while (b$'$) $\Rightarrow$ (b) and (c$'$) $\Rightarrow$ (c) are both trivial. The

argument for (c) $\Rightarrow$ (a) is identical to the proof of (4.1)(c) $\Rightarrow$ (a) (except it

uses Proposition (1.4)(3) in place of (1.3)(3).) (This argument also

shows that the paranthetical statements in (c) and (c$'$) hold.)

(a) $\Rightarrow$ (c$'$) This is very similar to the proof of (4.1)(a) $\Rightarrow$ (c). Suppose

that (a) holds, and let Q and k be as in (c$'$). By (1.1)(a), we may assume

that R is local at Q (and write $\overline{Q^k}$ instead of $Q^{<k>}$). Now (1.9)(a)

shows that (a) holds for S and IR^*. Since $\overline{I^m\ R_S} \cap R \subseteq$

$\overline{(IR^*)^m\ R_S^*} \cap R^*$, and since $\overline{Q^{*k}} \cap R = \overline{Q^k}$ [M2, Lemma 3.15], we

may assume that R is complete. Let q be any minimal prime of R, and

let p be a prime minimal over I + q. By (1.1)(b), $p \in \overline{Q^*}(I)$, so that (a)

shows S is disjoint from p, and hence from I + q. Let N be the nilradical

of R, so that N_S is the nilradical of R_S. As $I_S + q_S \neq R_S$ for all minimal

primes q of R, [M2, Lemma 3.11] shows that for $m \geq 1$, $\cap\ \overline{I^m\ R_S} = N_S$.

As S is disjoint from every minimal prime q of R, $N_S \cap R = N$, and so

$\cap (\overline{I^m\ R_S} \cap R) = N$. In the complete local ring R/N, we have

$\cap ((\overline{I^m\ R_S} \cap R)/N) = 0$. By [N3, 30.1], there is an $m \geq 1$ with

$(\overline{I^m\ R_S} \cap R)/N \subseteq (Q/N)^k = (Q^k + N)/N$. Therefore,

$\overline{I^m\ R_S} \cap R \subseteq Q^k + N \subseteq \overline{Q^k}$.

Proof of (4.4). (c$'$) $\Rightarrow$ (c) $\Rightarrow$ (b) and (c$'$) $\Rightarrow$ (b$'$) $\Rightarrow$ (b) are all trivial, and

(a) $\Rightarrow$ (c$'$) follows immediately from (1.2).

(b) $\Rightarrow$ (a) For $k \geq 1$, $\overline{u^k R(I)}$ has the form $\Sigma L_i t^i$ with $L_i = \overline{I^{k+i}} \cap I^i$.

Also, $\overline{u^{k+h} R(I)_S} \cap R(I)$ has the form $\Sigma J_i t^i$ with $J_i =$

$(\overline{I^{k+h+i} R_S} \cap R) \cap I^i$. If (b) holds, then $\overline{u^{k+h} R(I)_S} \cap R(I) \subseteq \overline{u^k R(I)}$.

Thus, (4.3)(b) holds for S and the ideal uR(I) (with m = k + h).

By (4.3)(a), S is disjoint from all the primes in $\overline{Q^*}(uR(I))$. Contracting

to R, S is disjoint from all primes in $\overline{A^*}(I)$, giving (a).

(4.6) Remark. Let b be a regular element of R, and suppose that $b^m R_S \cap R \subseteq bR$ for some m. Then by induction, $b^{m+k-1} R_S \cap R \subseteq bR^k$ for all $k \geq 1$. The proof of (4.2) makes use of that fact. However, in the proof of (4.4)(a) $\Rightarrow$ (c′), at the place we would like to use an similar fact for integral closures, we instead were forced to use (1.2), thus breaking the symmetry between the proofs of these two results. We now give an example showing that symmetry must be broken.

(4.7) Example. We will exhibit a ring R, a regular $b \in R$, and an S, such that $bR_S \cap R \subseteq \overline{bR}$, but such that for some $k \geq 1$, $b^n R_S \cap R \not\subseteq \overline{b^k R}$ for all $n \geq 1$. We use the construction of [N3, example 2, pp. 203-205]. This gives a 2-dimensional local domain (R, M) such that R′ is a finite R-module, and has exactly two maximal ideals N_1 and N_2, with height $N_1 = 1$ and height $N_2 = 2$, and with $M = N_1 \cap N_2$. Also, there is an $\alpha \in R'$ with $R' = R + R\alpha$, and with $N_1 = \alpha R'$. Let $\beta \in N_2 - N_1$, and let $b = \alpha\beta$. Note that $b \in N_1 \cap N_2 = M \subseteq R$. We claim that $\overline{bR} = \beta R' \cap R$. Clearly $\overline{bR} = bR' \cap R = \alpha\beta R' \cap R \subseteq \beta R' \cap R$. For the reverse inclusion, suppose $\gamma \in R'$ and $\beta\gamma \in \beta R' \cap R$. We want $\beta\gamma \in \overline{bR}$. Write $\gamma = x + y\alpha$ with x, y $\in$ R. Since $\beta y\alpha = yb \in \overline{bR}$, we need $\beta x \in \overline{bR}$. Now $\beta \in N_2$ implies $\beta\gamma$ and $\beta y\alpha = yb$ are in $N_2 \cap R = M$. Thus $\beta x \in M \subseteq N_1$. As $\beta \notin N_1$, $x \in N_1 = \alpha R'$. Thus $\beta x \in \beta\alpha R' \cap R = \overline{bR}$, as desired. This proves the claim. Now let $S' = R' - \cup\{p \in A^*(\beta R')\}$. Clearly $\beta R'_{S'} \cap R' = \beta R'$. Since $\beta \notin N_1 = \alpha R'$, we easily see that $\alpha \in S'$. Thus $\alpha\beta R'_{S'} = \beta R'_{S'}$, and so $bR'_{S'} \cap R' = \beta R'$. Let $S = S' \cap R$. We see that $bR_S \cap R \subseteq (bR'_{S'} \cap R') \cap R = \beta R' \cap R = \overline{bR}$, using the above claim. Thus we have found R, b regular in R, and S with $bR_S \cap R \subseteq \overline{bR}$.

Since $\beta \notin N_1$, $N_1 \notin A^*(\beta R')$. Since R′ is a Krull domain, primes in $A^*(\beta R')$ have height 1. Thus, $N_2 \notin A^*(\beta R')$, and we may pick $x \in N_1 \cap N_2$ with $x \in R' - \cup\{p \in A^*(\beta R')\} = S'$. As $N_1 \cap N_2 = M$,

$x \in S' \cap M = S \cap M$. Since height $N_1 = 1$ and since $b = \alpha\beta \in N_1$,

$N_1 \in \overline{A}^*(bR')$. By (1.13)(c), $M = N_1 \cap R \in \overline{A}^*(bR)$. Since $x \in S \cap M$, we

see that $S \not\subseteq R - \cup\{P \in \overline{A}^*(bR)\}$. It now follows from (4.3)(b) $\Rightarrow$ (a), (and

the fact that $\overline{A}^*(bR) = \overline{Q}^*(bR)$) that for some $k \geq 1$, there is no $n \geq 1$ such

that $b^n R_S \cap R \subseteq \overline{b^k} R$.

(4.8) Remark. Schenzel's theorems lead to nice alternate proofs of

$E(I) \subseteq A^*(I)$, $Q(I) \subseteq E(I)$ and $\overline{Q}^*(I) \subseteq \overline{A}^*(I)$. We present the first of
these, leaving the other two to the reader. Let $h \geq 0$ be large enough that

for all $k \geq 1$, Ass $R/I^{k+h} = A^*(I)$. Let $S = R - \cup\{Q \in A^*(I)\}$. Then for all

$k \geq 1$, S consists of regular elements modulo I^{k+h}. Thus, $I^{k+h} R_S \cap R =$

$I^{k+h} \subseteq I^k$. Now (4.2)(b) $\Rightarrow$ (a) shows that $S \subseteq R - \cup\{P \in E(I)\}$. This

shows that $\cup\{P \in E(I)\} \subseteq \cup\{Q \in A^*(I)\}$. However, $A^*(I)$ is finite, and so

we see that for any $P \in E(I)$, there is a $Q \in A^*(I)$ with $P \subseteq Q$. Now pick

$P \in E(I)$. Then $P_P \in E(I_P)$, and by what we have just shown applied in

R_P, there is a $Q_P \in A^*(I_P)$ with $P_P \subseteq Q_P$. As P_P is maximal in R_P, we

have $P_P = Q_P \in A^*(I_P)$, so that $P \in A^*(I)$. (To show $Q(I) \subseteq E(I)$,

compare (4.2) to (4.1), and to show $\overline{Q}^*(I) \subseteq \overline{A}^*(I)$, compare (4.4) to (4.3).)

Most of the material in this chapter comes from [M5].
Related work appears in the papers of Huckaba, [H], Ratliff, [R4], [R5],
and Verma, [V1], [V2].

5 THE RELATIVE REES RING OF I AND J

Notation. Let I and J be ideals of R. Recall that $R(I) = R[u, It]$. Let $R(I, J)$ denote the graded subring of $R[u, t]$ having the form $R(I, J) = \cdots + Rt^{-2} + Rt^{-1} + R + (I : <J>)t + (I^2 : <J>)t^2 + \cdots$.

In this brief chapter, we will investigate when $R(I, J)$ is either a finite module extension or an integral extension of $R(I)$.

Definitions. $\mathbf{P}(R) = \{P \in \text{Spec } R \mid P \in \text{Ass } R/bR \text{ for some } b \in P \text{ with }$ the image of b in Rp a regular element in Rp}. $\mathbf{E}(R) = \{P \in \text{Spec } R \mid$ $P \in E(bR) \text{ for some } b \in P \text{ with the image of b in Rp a regular element}$ in Rp}. $\mathbf{A}(R) = \{P \in \text{Spec } R \mid P \in \overline{A}^*(bR) \text{ for some } b \in P \text{ with the}$ image of b in Rp not in any minimal prime of Rp}.

(5.1) Lemma. Let $P \in$ Spec R. The following are equivalent.

i) $P \in \mathbf{E}(R)$ (respectively, $P \in \mathbf{A}(R)$).

ii) Essential grade Pp = 1 (respectively, asymptotic grade Pp = 1).

iii) $P \notin$ Ass R and the completion Rp* has a depth 1 prime divisor of zero (respectively, P is not minimal and Rp* has a depth 1 minimal prime).

Proof. We leave the easy proof of (i) $\Leftrightarrow$ (ii) to the reader. (Remark:
Note also that $P \in \mathbf{P}(R)$ if and only if grade $Pp = 1$.) For (i) $\Leftrightarrow$ (iii), we
do the essential case, the asymptotic case being similar. Say $P \in \mathbf{E}(R)$.
Then there is a $b \in R$ with $P \in \mathbf{E}(bR)$ and b regular in Rp. Now
$Pp \in \mathbf{E}(bRp)$, and since b is an essential sequence in Rp, (3.14) shows
that $Pp \in \mathbf{Q}(bRp)$. Thus Rp^* has a prime divisor of zero, q, such that
Pp^* is minimal over $bRp^* + q$. Since b is regular in Rp and in Rp^*, we
see that $P \notin \mathrm{Ass}\, R$ and that depth $q = 1$. Thus (i) $\Rightarrow$ (iii). The converse is
similar.

Definition. Let K be a regular ideal of R. The ideal transform of K is
$T(K) = \{x \mid x$ is in the total quotient ring of R, and $xK^n \subseteq R$ for some
integer $n \geq 1\}$. (Clearly $T(K)$ is a ring between R and its total quotient
ring.)

(5.2) Proposition. Let K be a regular ideal of R. Then

1) $T(K) = R$ if and only if $K \nsubseteq \cup\{P \in \mathbf{P}(R)\}$.

2) $T(K)$ is a finite R-module if and only if $K \nsubseteq \cup\{P \in \mathbf{E}(R)\}$.

3) $T(K)$ is integral over R if and only if $K \nsubseteq \cup\{P \in \mathbf{A}(R)\}$.

Proof. The proof of (1) is by standard arguments. (2) is by
[M2, Propositions 10.9 and 10.11(i) $\Leftrightarrow$ (ii)] and (5.1). (3) is by
[M2, Corollary 10.4] and (5.1).

(5.3) Theorem Let notation be as above. Then,

a) $R(I, J)$ is a finite $R(I)$-module if and only if $J \nsubseteq \cup\{P \in \mathbf{E}(I)\}$.

b) $R(I, J)$ is integral over $R(I)$ if and only if $J \nsubseteq \cup\{P \in \overline{\mathbf{A}}^*(I)\}$.

Proof. We prove (a), (b) being argued analogously. First, we claim that $R(I, J)$ equals the ideal transform $T((J, u)R(I))$. It is easily seen that $R(I, J)$ consists of those elements w of the total quotient ring of $R(I)$ such that $(R(I) : w)$ contains both a power of J and a power of u. These are exactly the elements w of the total quotient ring of $R(I)$ such that $(R(I) : w)$ contains a power of $(J, u)R(I)$. However, this also describes $T((J, u)R(I))$, proving the claim.

By (5.2)(2), $R(I, J)$ is a finite $R(I)$-module if and only if $(J, u)R(I) \not\subseteq \cup\{Q \in \mathbf{E}(R(I))\}$. We need to show that this is equvalent to having $J \not\subseteq \cup\{P \in \mathbf{E}(I)\}$. Suppose $(J, u)R(I) \subseteq Q \in \mathbf{E}(R(I))$. By (5.1)(iii), $R(I)_Q^*$ has a depth 1 prime divisor of zero, say q. Now u is regular in $R(I)$, and so we must have that QQ^* is minimal over $uR(I)_Q^* + q$. By definition, $Q \in \mathbf{Q}(uR(I))$, so that $Q \cap R \in \mathbf{E}(I)$.

As $J \subseteq Q \cap R$, we see that $J \subseteq \cup\{P \in \mathbf{E}(I)\}$. This proves one direction of (a). The other direction is argued similarly.

We wish to relate (5.3) to some results in the previous chapter. To do so, we need to translate the notation of the previous chapter into the notation of this chapter.

(5.4) Lemma. (i) Given any ideal J of R, there is a multiplicatively closed set S of R such that $I^n : <J> = I^nR_S \cap R$ and $\overline{I^n R_S} \cap R = \overline{I^n} : <J>$ for all $n \geq 1$, and such that if $A(I)$ is any subset of $\cup Ass\ R/I^n$, $n \geq 1$, then $J \not\subseteq \cup\{P \in A(I)\}$ if and only if $S \subseteq R - \cup\{P \in A(I)\}$.

(ii) Conversely, given any multiplicatively closed subset S of R, there is an ideal J of R such that $I^nR_S \cap R = I^n : <J>$ and $\overline{I^n R_S} \cap R = \overline{I^n} : <J>$ for all $n \geq 1$, and such that if $A(I)$ is any subset of $\cup Ass\ R/I^n$, $n \geq 1$, then $S \subseteq R - \cup\{P \in A(I)\}$ if and only if $J \not\subseteq \cup\{P \in A(I)\}$.

Proof. (i) Given J, pick $x \in J$ such that x is in a prime in $\cup Ass\ R/I^n$ over $n \geq 1$ if and only if J is contained in that prime. (This can be done, since $\cup Ass\ R/I^n$ is finite, being at worst a little bigger than $A^*(I)$.) Let

$S = \{x, x^2, x^3, \cdots\}$. Now for any $n \geq 1$, any prime divisor of either I^n or $\overline{I^n}$ is in $\cup \text{Ass } R/I^n$, (the second case using (1.2) and $\overline{A}^*(I) \subseteq A^*(I)$). Therefore, using (1.7)(a), it is not hard to see that $I^n : <J> = I^n R_S \cap R$ and $\overline{I^n} : <J> = \overline{I^n} R_S \cap R$ for all $n \geq 1$. This proves the first part of (i). Now J is contained in a prime in $\cup \text{Ass } R/I^n$, $n \geq 1$, if and only if S intersects that prime nontrivially. Since $A(I) \subseteq \cup \text{Ass } R/I^n$, $n \geq 1$, the second part of (i) follows. (Note: this construction was used in proving (1.7).)

(ii) Given S, pick $y \in S$ such that y is in a prime in $\cup \text{Ass } R/I^n$, $n \geq 1$, if and only if S meets that prime nontrivially. Let $J = yR$. As in (i), it is not hard to see that $I^n R_S \cap R = I^n : <J>$ and $\overline{I^n} R_S \cap R = \overline{I^n} : <J>$ for all $n \geq 1$, giving the first part of (ii). Also, a prime in $\cup \text{Ass } R/I^n$, $n \geq 1$, contains J if and only if it meets S nontrivially, and the rest of (ii) follows.

(5.5) Remark. Using (5.4), it is easy to translate (4.1), (4.2), (4.3), and (4.4) into a form involving $I^n : <J>$. For example, (4.2)(a) $\Leftrightarrow$ (b) becomes $J \not\subseteq \cup \{P \in E(I)\}$ if and only if there is an $h \geq 0$ such that for all $k \geq 1$, $I^{k+h} : <J> \subseteq I^k$. Here we used (5.4) with $A(I) = E(I)$. We leave the rest of these translations to the reader. (These are the forms which Schenzel used in [Sc].)

(5.6) Remark. We give another proof of (5.3)(a). By (4.2)(a)$\Leftrightarrow$(b) (as translated in (5.5)) $J \not\subseteq \cup \{P \in E(I)\}$ if and only if there is an $h \geq 0$ such that for all $k \geq 1$, $I^{k+h} : <J> \subseteq I^k$. However, it is clear that such an h exists if and only if $u^h R(I, J) \subseteq R(I)$, which in turn is clearly equivalent to having $R(I, J)$ a finite $R(I)$-module, since $R(I, J) \subseteq R(I)[u^{-1}]$. (Obviously, we could have argued the other way around, starting with (5.3)(a), and giving a new proof of (4.2)(a)$\Leftrightarrow$(b).)

(5.7) Remark. We give another proof of (5.3)(b). If we use (5.4) to translate (4.4)(a)$\Leftrightarrow$(c), we find that $J \not\subseteq \cup\{P \in \overline{A}^*(I)\}$ if and only if for all $k \geq 1$, $I^k : \langle J \rangle \subseteq \overline{I^k}$. This is equivalent to saying that

$$R(I, J) \subseteq R[\overline{I}\, t,\ \overline{I^2}\, t^2,\ \overline{I^3}\, t^3, \dots] = R(I)' \cap R[u, u^{-1}].$$ Clearly this occurs exactly when $R(I, J)$ is integral over $R(I)$. (Obviously, we could have argued the other way around, starting with (5.3)(b), and giving a new proof of (4.4)(a)$\Leftrightarrow$(c).)

    This material is taken from [M4], which was motivated by work in [Sc].

6 TWO ASYMPTOTIC FUNCTIONS

Notation. I and J will be ideals of R. $R(I, J)$ will be as in Chapter 5. For $n \geq 0$, by $R(I, J)_n$ we will mean the $R(I)$-submodule of $R(I, J)$ having the form $R(I, J)_n = \cdots + Rt^{-2} + Rt^{-1} + R + (I : J^n)t + (I^2 : J^n)t^2 + \cdots$. (Note that since $I^m : \langle J \rangle = \cup(I^m : J^n)$, $n \geq 1$, $R(I, J) = \cup R(I, J)_n$, over $n \geq 0$. Also note that $R(I, J)_0 = R(I)$.)

In this chapter, we will use $R(I, J)$ to study two interesting functions. We will often need to assume that $R(I, J)_n$ is a finite $R(I)$-module for all $n \geq 1$. This is a mild assumption. For example, we now show that it holds if J is a regular ideal of R, and is equivalent to J being regular if R is local.

(6.1) Lemma. a) If J is regular, $R(I, J)_n$ is a finite $R(I)$-module for all $n \geq 1$.

b) If for some $n \geq 1$, $R(I, J)_n$ is a finite $R(I)$-module, then for $k \geq 1$, $(\cap I^k : J) = \cap I^k$.

c) If R is local, $R(I, J)_n$ is a finite $R(I)$-module (for any $n \geq 1$) if and only if J is regular.

Proof. Certainly $J^n R(I, J)_n \subseteq R(I)$. If x is a regular element of J, then $R(I, J)_n \subseteq R(I)x^{-n}$, proving part (a). Now suppose $n \geq 1$ and $R(I, J)_n$ is a finite R-module. Then for a sufficiently large integer h, $u^h R(I, J)_n \subseteq R(I)$. It easily follows that for all $k \geq 1$, $I^{h+k} : J^n \subseteq I^k$. Thus $(\cap I^k : J) \subseteq (\cap I^k : J^n) \subseteq (\cap I^{k+h} : J^n) \subseteq \cap I^k$, so (b) holds. Now (c) follows from (a) and (b) and the fact that in a local ring, $\cap I^k = 0$.

Notation. For $m \geq 0$, the chain $(I^m : J^0) \subseteq (I^m : J^1) \subseteq (I^m : J^2) \subseteq \ldots$ eventually stabilizes to $I^m : \langle J \rangle$. We let $\alpha(I, J, m) = \alpha(m)$ be the least nonnegative integer such that $(I^m : J^{\alpha(m)}) = I^m : \langle J \rangle$. (We will only use the notation $\alpha(I, J, m)$ when it is useful to exhibit which ideals I and J are under consideration.)

(6.2) Theorem. a) The function $\alpha(m)$ is eventually nondecreasing.

b) $\alpha(m) \leq n$ (an integer) for all $m \geq 0$ if and only if $R(I, J)_n = R(I, J)$.

c) $J \not\subseteq \cup \{P \in E(I)\}$ if and only if $R(I, J)_n$ is a finite $R(I)$-module for all $n \geq 1$ and $\alpha(m)$ is eventually constant. In particular, if J is regular, then $J \not\subseteq \cup \{P \in E(I)\}$ if and only if $\alpha(m)$ is eventually constant.

Proof. a) It follows easily from [M2, Lemma 1.1(b)], there is an integer $a \geq 0$ such that $(I^{m+j} : \dot{J}) \cap I^a = I^m$ for all $m \geq a$ and $j \geq 0$. Let $n = \min \{\alpha(m) \mid m \geq a\}$, and suppose that $n = \alpha(k)$ with $k \geq a$. We will show that our function is nondecreasing for $m \geq k$. Let $k \leq r < s$. We must show that $\alpha(r) \leq \alpha(s)$. Clearly $I^{s-r}(I^r : \langle J \rangle) \subseteq I^s : \langle J \rangle = (I^s : J^{\alpha(s)})$, the last equality by the definition of $\alpha(s)$. Thus, $J^{\alpha(s)}(I^r : \langle J \rangle) \subseteq (I^s : I^{s-r})$. We also claim that $J^{\alpha(s)}(I^r : \langle J \rangle) \subseteq I^a$, deferring the proof momentarily. The opening sentence of this proof shows that $J^{\alpha(s)}(I^r : \langle J \rangle) \subseteq I^r$. That is, $I^r : \langle J \rangle \subseteq (I^r : J^{\alpha(s)})$. The reverse inclusion is automatic, and so the definition shows that $\alpha(r) \leq \alpha(s)$, as desired. It remains to prove our claim. We have $s > r \geq k \geq a$, and (by the choice of n) $\alpha(s) \geq n = \alpha(k)$. Together, these give that $I^r : \langle J \rangle \subseteq I^k : \langle J \rangle = (I^k : J^{\alpha(k)}) \subseteq (I^k : J^{\alpha(s)})$. Thus, $J^{\alpha(s)}(I^r : \langle J \rangle) \subseteq I^k \subseteq I^a$, proving the claim.

b) $R(I, J) = R(I, J)_n$ if and only if $I^m : \langle J \rangle = (I^m : J^n)$ for all $m \geq 0$ if and only if $\alpha(m) \leq n$ for all $m \geq 0$.

c) Suppose that $J \not\subseteq \cup \{P \in E(I)\}$. By (5.3)(a), $R(I, J)$ is a finite $R(I)$-module. Thus each submodule $R(I, J)_n$, $n \geq 1$, is a finite $R(I)$-module. Also, $R(I, J)$ is the union of the increasing chain $R(I, J)_n$, $n = 1, 2, 3, \cdots$, and so $R(I, J) = R(I, J)_n$ for some $n \geq 1$. By (b), $\alpha(m) \leq n$ for all $m \geq 1$. By (a), $\alpha(m)$ is eventually constant. (Alternate argument.

Take a finite set of homogeneous generators for $R(I, J)$ over $R(I)$, and let c be the highest of their degrees. Let $n = \max \{\alpha(1), \ldots, \alpha(c)\}$. Then J^n sends each generator into $R(I)$, so $J^n R(I, J) \subseteq R(I)$. Therefore, for any $m \geq 0$, $J^n(I^m : <J>) \subseteq I^m$, so that $(I^m : <J>) = (I^m : J^n)$. This shows that $\alpha(m) \leq n$ for all $m \geq 1$, so use (a).)

Conversely, suppose that $R(I, J)_n$ is a finite $R(I)$-module for all $n \geq 1$ and that $\alpha(m)$ is eventually constant. Then by (a), we have that $\alpha(m) \leq n$ for some $n \geq 1$ and all $m \geq 0$. By (b), $R(I, J) = R(I, J)_n$.

Therefore, $R(I, J)$ is a finite $R(I)$-module. By (5.3)(a), $J \not\subseteq \cup \{P \in E(I)\}$.

The final sentence in (c) follows from the first part of (c) together with (6.1)(a).

(6.3) Lemma. Let $n \geq 0$ be a fixed integer. The following are equivalent for an integer $h \geq 0$.

1) $(I^{h+r} : J^n) \subseteq I^r$ for all $r \geq 1$.

2) $(u^{h+r}R(I) : J^n R(I)) = u^r(u^h R(I) : J^n R(I))$ for all $r > 1$.

3) $u^h R(I, J)_n = (u^h R(I) : J^n R(I))$.

4) $u^h R(I, J)_n \subseteq R(I)$.

Furthermore, $R(I, J)_n$ is a finite $R(I)$-module if and only if there is an $h \geq 0$ for which these statements are all true.

Proof. 1) $\Rightarrow$ 3) It is easily seen that $u^h R(I, J)_n = \Sigma(I^{r+h} : J^n)t^r$, while $(u^h R(I) : J^n R(I)) = \Sigma((I^{r+h} : J^n) \cap I^r)t^r$, both sums over all integers r (negative powers of I being R). If (1) holds, then $(I^{r+h} : J^n) \subseteq I^r$ for all r, (the cases $r \leq 0$ being trivial). Thus $(u^h R(I) : J^n R(I)) = \Sigma(I^{r+h} : J^n)t^r = u^h R(I, J)_n$, and so (3) holds.

3) $\Rightarrow$ 4) $\Rightarrow$ 1) The first is trivial, since $(u^h R(I) : J^n R(I)) \subseteq R(I)$. If (4) holds, then $u^h R(I, J)_n = \Sigma(I^{r+h} : J^n)t^r$ is contained in $R(I) = \Sigma I^r t^r$. Clearly this implies (1).

$4) \Rightarrow 2)$ Suppose (4) holds, so that $u^h R(I, J)_n \subseteq R(I)$. Certainly for any

$r \geq 1$, we also have $u^{h+r} R(I, J)_n \subseteq R(I)$. We now apply $(4) \Rightarrow (3)$ (which

is already proved) to both h and h + r. We see that $u^h R(I, J)_n =$

$(u^h R(I) : J^n R(I))$, and $u^{h+r} R(I, J)_n = (u^{h+r} R(I) : J^n R(I))$. This shows

that (2) holds.

$2) \Rightarrow 1)$ If (2) holds, then intersecting both sides of that equation with R

shows that (1) holds.

Finally, it is easily seen that (1) holds if and only if $u^h R(I, J)_n \subseteq R(I)$,

and that this is true for some $h \geq 0$ if and only if $R(I, J)_n$ is a finite

R(I)-module.

Definition. Suppose that $R(I, J))_n$ is a finite R(I)-module for all $n \geq 0$.

For each $n \geq 0$, let $\beta(I, J, n) = \beta(n)$ be the smallest integer h for which the

equivalent statements in (6.3) hold. (We will only use the notation

$\beta(I, J, n)$ when it is useful to exhibit which ideals I and J are under

consideration.)

(6.4) Theorem. Suppose that $R(I, J)_n$ is a finite R(I)-module for all $n \geq 0$

(so that the function $\beta(n)$ is defined). Then $\beta(n)$ is nondecreasing.

Also $\beta(n)$ is eventually constant if and only if $J \not\subseteq \cup\{P \in E(I)\}$.

Proof. That $\beta(n)$ is nondecreasing is immediate from the definition,

(6.3)(4), and the fact that $R(I, J)_n$, $n \geq 0$, is an increasing chain of

modules. Suppose that $\beta(n)$ is eventually constant, say equaling c.

Then $\beta(n)$ is bounded by c, and so by (6.3)(4), we have $u^c R(I, J)_n \subseteq R(I)$

for all $n \geq 0$. As R(I. J) is the union of all $R(I, J)_n$, we get $u^c R(I, J) \subseteq$

R(I). This shows that R(I, J) is a finite R(I)-module, and (5.3)(a) gives

$J \not\subseteq \cup\{P \in E(I)\}$. Conversely, if $J \not\subseteq \cup\{P \in E(I)\}$, then R(I, J) is a finite

R(I)-module. Therefore, for large n we have $R(I, J)_n = R(I, J)_{n+1} = \cdots$.

Thus, by (6.3)(4), $\beta(n) = \beta(n + 1) = \cdots$. (Remark: This argument,

together with (6.2)(b), shows that if $\alpha(m) \leq n$ for all $m \geq 0$, then

$\beta(n) = \beta(n+1) = \cdots$.)

Despite the title of this chapter, we will look briefly at a third function, which the next lemma allows us to define.

(6.5) Lemma. Let A be a graded R(I)-module with $R(I) \subseteq A \subseteq R[u, t]$, having the form $A = \cdots + Rt^{-1} + R + K_1 t + K_2 t^2 + \cdots$. Then A is a finite R(I)-module if and only if for all large integers h, $K_{r+h} = I^r K_h$ for all $r \geq 1$. If A is a finite R(I)-module, then for all large h, $K_h{}^r = K_{hr}$ for all $r \geq 1$.

Proof. Suppose that A is a finite R(I)-module. Pick a finite set of homogeneous module generators for A over R(I). We may assume that 1 is in this set, and that the other homogeneous elements in this set have positive degrees. Let h be the largest of those degrees. For any $r \geq 1$, it is easy to see that $K_{r+h} = \Sigma I^{r+i} K_{h-i}$ over i = 0, 1, ..., h. Since for $0 \leq i \leq h$,

$I^{r+i} K_{h-i} \subseteq I^r K_h$, we get $K_{r+h} = I^r K_h$, proving one direction of the first part. The other direction is easy. If A is a finite R-module, then by what we have just proved, for large h and all $r \geq 1$, $K_{hr} = K_{h(r-1)+h} =$

$I^{h(r-1)} K_h \subseteq K_h{}^{r-1} K_h = K_h{}^r \subseteq K_{hr}$, which proves the second part.

Definition. Suppose that $R(I, J)_n$ is a finite R(I)-module for all $n \geq 0$. For $n \geq 0$, (6.5) shows that there is an $h \geq 0$ such that $(I^{r+h} : J^n) = I^r(I^h : J^n)$ for all $r \geq 1$. Define $\gamma(n)$ to be the least such h.

We now state everything we know about $\gamma(n)$.

(6.6) Theorem. Suppose that $R(I, J)_n$ is a finite R(I)-module for all $n \geq 0$ (so that $\gamma(n)$ exists). Then the following are equivalent.

i) $\gamma(n)$ is eventually constant.

ii) $\{\gamma(n) \mid n \geq 0\}$ is bounded above.

iii) $J \not\subseteq \cup \{P \in E(I)\}$.

Proof. (i) $\Rightarrow$ (ii) is trivial. Suppose that (ii) holds. Since for any $n \geq 0$ and $r \geq 1$, we have $(I^{\gamma(n)+r} : J^n) = I^r(I^{\gamma(n)} : J^n)$, clearly $(I^{\gamma(n)+r} : J^n) \subseteq I^r$. By the definition of $\beta(n)$ (using (6.3)(1)), we have $\beta(n) \leq \gamma(n)$. By (ii), we see that $\beta(n)$ is bounded above. Now (6.4) shows that $\beta(n)$ is eventually constant and also shows that (iii) holds. Finally, suppose (iii) holds. By (5.3)(a), $R(I, J)$ is a finite $R(I)$-module. Thus for large enough n, $R(I, J)_n = R(I, J)_{n+1} = \cdots = R(I, J)$. Thus, for any integer k, we have $(I^k : J^n) = (I^k : J^{n+1}) = \cdots$. It easily follows that $\gamma(n) = \gamma(n + 1) = \cdots$, showing that (i) holds.

In the remainder of this chapter, we will investigate the behavior of $\alpha(m)/m$ as $m \to \infty$, and $\beta(n)/n$ as $n \to \infty$. We will consider three topics. The first is whether $\alpha(m)/m$ and $\beta(n)/n$ have limits as their arguments go to infinity. (Whenever we discuss $\beta(n)$, we will assume that J is regular, so that by (6.1)(a), $R(I, J)_n$ will be a finite $R(I)$-module for all $n \geq 1$, so that $\beta(n)$ will exist.)

(6.7) Proposition. Let J be regular. Then lim $\beta(I, J, n)/n$ exists as $n \to \infty$, and is equal to or less than $\beta(I, J, 1)$.

Proof. Let n and m be positive integers. We claim that $\beta(n + m) \leq \beta(n) + \beta(m)$. By the definition, we have $(I^{\beta(n)+r} : J^n) \subseteq I^r$ and $(I^{\beta(m)+r} : J^m) \subseteq I^r$ for all $r \geq 1$. Now $(I^{\beta(n)+\beta(m)+r} : J^{n+m}) = ((I^{\beta(n)+\beta(m)+r} : J^n) : J^m) \subseteq (I^{\beta(m)+r} : J^m) \subseteq I^r$. The minimality of $\beta(n + m)$ now shows that our claim is true.

Let $B = \lim \inf \{\beta(n)/n \mid n = 1, 2, 3, \cdots\}$. For any $\varepsilon > 0$, there is a sufficiently large value of m with $\beta(m)/m < B + \varepsilon/2$. Fix m. For any $n \geq 1$, write $n = qm + r$ with q and r nonnegative integers and $r < m$. From our earlier claim, $\beta(n) = \beta(qm + r) \leq q\beta(m) + r\beta(1) < qm(B + \varepsilon/2) + m\beta(1)$. Thus $\beta(n)/n < (qm/n)(B + \varepsilon/2) + (m/n)\beta(1) \leq (B + \varepsilon/2) + \beta(1)/q$. Since as $n \to \infty$, clearly $q \to \infty$, we see that for large n, $\beta(1)/q < \varepsilon/2$. Thus $\beta(n)/n \leq B + \varepsilon$ for all large n. Since B is the lim inf of our sequence, we now see that B is in fact the limit of $\beta(n)/n$ as $n \to \infty$. Finally, since for all $n \geq 1$, the claim of the previous paragraph shows that $\beta(n) \leq n\beta(1)$, we have $\beta(n)/n \leq \beta(1)$, which proves the final statement of the result.

The behavior of $\alpha(I, J, m)/m$ as $m \to \infty$ is not as easy to discern. We can treat the case that I is a principal regular ideal. (6.8) and (6.9) are due to T. Marley.

(6.8) Proposition. Let $I = bR$ be a regular ideal. As $m \to \infty$, $\lim \alpha(bR, J, m)/m$ exists, and is equal to or less then $\alpha(bR, J, 1)$.

Proof. We claim that for positive integers m and n, $\alpha(m + n) \leq \alpha(m) + \alpha(n)$. It will suffice to show that $(b^{m+n}R : J^{\alpha(m)+\alpha(n)+1}) \subseteq (b^{m+n}R : J^{\alpha(m)+\alpha(n)})$, since then equality holds between these two ideals, and the minimality of $\alpha(m + n)$ will imply the claim. Let $x \in (b^{m+n}R : J^{\alpha(m)+\alpha(n)+1})$. Then $xJ^{\alpha(m)+\alpha(n)+1} \subseteq b^{m+n}R \subseteq b^mR$. The definition of $\alpha(m)$ shows that $xJ^{\alpha(m)} \subseteq b^mR$, and so if $H = (xJ^{\alpha(m)} : b^mR)$, then $xJ^{\alpha(m)} = b^mH$. Thus $b^mHJ^{\alpha(n)+1} = xJ^{\alpha(m)+\alpha(n)+1} \subseteq b^{m+n}R$. As b is regular, $HJ^{\alpha(n)+1} \subseteq b^nR$. The definition of $\alpha(n)$ now shows that $HJ^{\alpha(n)} \subseteq b^nR$. Therefore, $xJ^{\alpha(m)+\alpha(n)} = b^mHJ^{\alpha(n)} \subseteq b^{m+n}R$, showing that $x \in (b^{m+n}R : J^{\alpha(m)+\alpha(n)})$, as desired. The rest of the argument is as in the proof of (6.7).

(6.9) Remark. It is easily seen that $(u^mR(I) : J^nR(I)) \cap R = (I^m : J^n)$. From this, it quickly follows that $\alpha(I, J, m) \leq \alpha(uR(I), JR(I), m) \leq m\,\alpha(uR(I), JR(I), 1)$, using the claim in the previous proof. Therefore, we see that for arbitrary ideals I and J, the set $\{\alpha(I, J, m)/m \mid m = 1, 2, 3, ...\}$ is bounded above by $\alpha(uR(I), JR(I), 1)$.

We now turn to our second topic. If $\beta(n)$ is eventually constant, then clearly $\lim \beta(n)/n = 0$ as $n \to \infty$. We will prove that the converse holds as well. We suspect that the corresponding converse for $\lim \alpha(I, J, m)/m$ also holds, but can only prove it when I and J are regular with I principal. To reach these goals, we must introduce a new function (again disregarding the chapter title).

Definition. Let J be regular. For $n \geq 1$ an integer, define $\delta(I, J, n)$ to be the least integer such that $(I^{\delta(I,J,n)+1} : J^n) \subseteq I$.

(6.10) Remarks: a) The definitions easily show that $0 \leq \delta(I, J, n) \leq \beta(I, J, n) \leq \gamma(I, J, n)$ for all $n \geq 1$. (In particular, $\delta(I, J, n)$ exists.) In general, $\delta(I, J, n) \neq \beta(I, J, n)$. In example (2.3), $M \in A^*(q)$. Thus for large n, $M \in \text{Ass } R/q^n$. Clearly $M \notin \text{Ass } R/q$. As $(q : M) = q$, the definiton shows that $\delta(q, M, 1) = 0$. However, $(q^{0+n} : M) \not\subseteq q^n$, so that $\beta(q, M, 1) \neq 0$.

b) Since $(I^{\delta(I,J,n+1)+1} : J^n) \subseteq (I^{\delta(I,J,n+1)+1} : J^{n+1}) \subseteq I$, clearly $\delta(I, J, n) \leq \delta(I, J, n + 1)$, so that $\delta(I, J, n)$ is nondecreasing.

We come to a key argument.

(6.11) Proposition. Let J be regular. Let A denote lim sup $\{\alpha(I, J, m)/m\}$ and D denote lim sup $\{\delta(I, J, n)/n\}$. Then either $AD \geq 1$, or $\delta(I, J, n)$ is eventually constant. (Note that A and D are finite. For A, use (6.9). For D, use (6.10)(a) and (6.7).)

Proof. Write $\alpha(m)$ for $\alpha(I, J, m)$ and $\delta(n)$ for $\delta(I, J, n)$. Suppose $AD < 1$. We will show that $\delta(n)$ is eventually constant. Since $A < 1/D$, let s be a rational number with $A < s < 1/D$. (Here, if $D = 0$, we let $1/D = \infty$.) By the definition of A, for all sufficiently large m, $\alpha(m)/m < s$, so that $\alpha(m) < ms$. As s is rational, we may choose a large integer m with both $\alpha(m) < ms$, and with ms an integer. Since $s < 1/D$, $D < 1/s$. Therefore, for large n, $\delta(n)/n < 1/s$, so that $s\delta(n) < n$. If for n we take ms, and let m be large enough, we now have ms is an integer, $\alpha(m) < ms$, and $\delta(ms) < m$ (since $s\delta(ms) < ms$). Since $\alpha(m) < ms$ and ms is an integer, by the definition of $\alpha(m)$ we see that $(I^m : J^{ms}) = (I^m : J^{ms+k})$ for all integers $k \geq 0$. However, since $m > \delta(ms)$, $(I^m : J^{ms}) \subseteq (I^{\delta(ms)+1} : J^{ms}) \subseteq I$, the second inclusion using the definition of $\delta(ms)$. Combining the last two sentences, we see that for all $k \geq 0$,

$(I^m : J^{ms+k}) \subseteq I$. By the definition of $\delta(ms + k)$, we see that for all $k \geq 0$, $\delta(ms + k) \leq m - 1$. Thus $\delta(n)$ is eventually bounded. Since (6.10)(b)) shows $\delta(n)$ is nondecreasing, it is eventually constant.

(6.12) Remark. It follows easily from (6.11) that as $n \to \infty$,

$\lim \delta(I, J, n)/n = 0$ if and only if $\delta(I, J, n)$ is eventually constant.

We can now quickly reach our goal.

(6.13) Lemma. Let $I = bR$ and J be regular ideals.
Then $\delta(bR, J, n) = \beta(bR, J, n) = \gamma(bR, J, n)$ for all $n \geq 1$.

Proof. By (6.10)(a), it will suffice to show that $\gamma(bR, J, n) \leq \delta(bR, J, n)$. This follows from the definitions, and the easily proved fact that if for some $h \geq 0$, and $n \geq 1$, $(b^{h+1}R : J^n) \subseteq bR$, then for all $r \geq 1$, $(b^{h+r}R : J^n) = b^r(b^hR : J^n)$.

(6.14) Lemma. Let J be regular. Then $\beta(I, J, n) = \beta(uR(I), JR(I), n)$ for all $n \geq 1$.

Proof. For any $m \geq 1$, $(u^mR(I) : J^nR(I)) = \Sigma[(I^{m+s} : J^n) \cap I^s]t^s$ over all integers s. Let $r \geq 1$ and let $m = \beta(I, J, n) + r$. Then

$(u^{\beta(I,J,n)+r}R(I) : J^nR(I)) = \Sigma[(I^{\beta(I,J,n)+r+s} : J^n) \cap I^s]t^s$. However, by

definition of $\beta(I, J, n)$, $(I^{\beta(I,J,n)+r+s} : J^n)$ is contained in I^{r+s}.

Therefore, we see that $(u^{\beta(I,J,n)+r}R(I) : J^nR(I)) \subseteq \Sigma I^{r+s}t^s = u^rR(I)$. The minimality of $\beta(uR(I), JR(I), n)$ now shows that

$\beta(uR(I), JR(I), n) \leq \beta(I, J, n)$. For the reverse, note that since by

definition $(u^{\beta(uR(I),JR(I),n)+r}R(I) : J^nR(I)) \subseteq u^rR(I)$ for all $r \geq 1$, the first sentence of this proof (with $s = 0$) shows that

$(I^{\beta(uR(I),JR(I),n)+r} : J^n) \subseteq I^r$. The minimality of $\beta(I, J, n)$ now

shows that $\beta(I, J, n) \leq \beta(uR(I), JR(I), n)$.

(6.15) Proposition. Let J be regular. As n $\to \infty$, lim $\beta(I, J, n)/n = 0$

if and only if $\beta(I, J, n)$ is eventually constant

if and only if $J \nsubseteq \cup\{P \in E(I)\}$.

Proof. The second equivalence is by (6.4). For the first equivalence, by (6.14), we may replace R, I, and J, by R(I), uR(I), and JR(I). Thus, we will assume that I = bR, with b a regular element of R. Now (6.13) and (6.12) give the first equivalence.

(6.16) Proposition. Let I = bR and J both be regular. Then as m $\to \infty$, lim $\alpha(bR, J, m)/m = 0$ if and only if $\alpha(bR, J, m)$ is eventually constant if and only if $J \nsubseteq \cup\{P \in E(bR)\}$.

Proof. If lim $\alpha(bR, J, m)/m = 0$, then in (6.11), A = 0, and that result shows that $\delta(bR, J, n)$ is eventually constant. By (6.13), $\beta(bR, J, n)$ is eventually constant. By (6.4), $J \nsubseteq \cup\{P \in E(bR)\}$. Using (6.2)(c), the rest is trivial.

(6.17) Corollary. Let J be regular. Then as m $\to \infty$, lim $\alpha(uR(I), JR(I) m)/m = 0$ if and only if $\alpha(uR(I), JR(I), m)$ is eventually constant if and only if $\alpha(I, J, m)$ is eventually constant if and only if $J \nsubseteq \cup\{P \in E(I)\}$.

Proof. The first equivalence is by (6.16). By (3.14), E(uR(I)) = Q(uR(I)), so that the definition of E(I) shows that E(I) = $\{Q \cap R \mid Q \in E(uR(I))\}$. Therefore, $J \nsubseteq \cup\{P \in E(I)\}$ if and only if $JR(I) \nsubseteq \cup\{Q \in E(uR(I))\}$. Combining this with (6.2)(c) gives the other equivalences.

(6.18) Question. With J regular, let A = lim sup $\{\alpha(I, J, m)/m\}$ and let B = lim $\beta(I, J, n)/n$. If AB < 1, is $\beta(I, J, n)$ eventually constant? (If true, we could let the I in (6.16) be arbitrary.)

The third topic, to which we now turn, is the most difficult of the three. We discuss some cases in which we know that the limits under consideration are rational. (We know of no cases when they are not rational.)

(6.19) Lemma. If b and c are regular elements of R,
$\alpha(bR, cR, m) = \beta(cR, bR, m)$ for all $m \geq 0$.

Proof. We will first show that $\alpha(bR, cR, m) \leq \beta(cR, bR, m)$. By the definition of $\alpha(bR, cR, m)$, it will suffice to show that

$(b^m R : {}_c\beta(cR,bR,m)+1 R) \subseteq (b^m R : {}_c\beta(cR,bR,m) R)$. Consider

$x \in (b^m R : {}_c\beta(cR,bR,m)+1 R)$. Write $xc^{\beta(cR,bR,m)+1} = yb^m$, for some

$y \in R$. Then $y \in (c^{\beta(cR,bR,m)+1} R : b^m R) \subseteq cR$, the last inclusion by

definition of $\beta(cR, bR, m)$. Write $y = cz$. As c is regular, our earlier

equation yields $xc^{\beta(cR,bR,m)} = zb^m$. Thus $x \in (b^m R : {}_c\beta(cR,bR,m) R)$.

This shows $(b^m R : {}_c\beta(cR,bR,m)+1 R) \subseteq (b^m R : {}_c\beta(cR,bR,m) R)$.
Therefore, $\alpha(bR, cR, m) \leq \beta(cR, bR, m)$.

For the reverse inequality, if we can show that

$(c^{\alpha(bR,cR,m)+r} R : b^m R) \subseteq c^r R$ for all $r \geq 1$, by the minimality of

$\beta(cR, bR, m)$ we will then have $\beta(cR, bR, m) \leq \alpha(bR, cR, m)$, and so will

be done. Suppose that $w \in (c^{\alpha(bR,cR,m)+r} R : b^m R)$, and write $wb^m = vc^{\alpha(bR,cR,m)+r}$ for some $v \in R$. Then $v \in (b^m R : c^{\alpha(bR,cR,m)+r} R) = (b^m R : c^{\alpha(bR,cR,m)} R)$, the equality coming easily from the definition of $\alpha(bR, cR, m)$. Write $vc^{\alpha(bR,cR,m)} = sb^m$ for some $s \in R$.
Combining this with an earlier equation shows $w = sc^r$. Thus
$(c^{\alpha(bR,cR,m)+r} R : b^m R) \subseteq c^r R$, as desired.

(6.20) Remark. Let $R = K[X, Y]$ with K a field and X and Y indeterminates. We easily see that $\alpha((X, Y)R, XR, m) = m = \beta((X, Y)R, XR, m)$, and $\alpha(XR, (X, Y)R, m) = 0 = \beta(XR, (X, Y)R, m)$. From this, it is easily seen that (6.19) fails if either I or J is not principal.

(6.21) Lemma. Let b and c be regular elements of R. Let $R \subseteq T$, where the ring T is a finite R-module. Then there is an integer $d \geq 0$ such that $\alpha(bR, cR, m) \leq \alpha(bT, cT, m) + d$ for all $m \geq 1$. If we also have $T \subseteq R_b$, then there is an integer $f \geq 0$ such that $\alpha(bT, cT, m) \leq \alpha(bR, cR, m) + f$.

In this case, as $m \to \infty$, $\lim \alpha(bR, cR, m)/m = \lim \alpha(bT, cT, m)/m$.

Proof. By the Artin-Rees lemma, we may choose d such that for all $n \geq d$, $c^n T \cap R = c^{n-d}(c^d T \cap R)$. By (6.19), it will suffice to show that $\beta(cR, bR, m) \leq \beta(cT, bT, m) + d$. For this, the minimality of $\beta(cR, bR, m)$ shows it will suffice to prove that

$(c^{\beta(cT,bT,m)+d+r}R : b^m R) \subseteq c^r R$ for all $r \geq 1$. Now

$(c^{\beta(cT,bT,m)+d+r}R : b^m R) \subseteq (c^{\beta(cT,bT,m)+d+r}T : b^m T) \subseteq c^{d+r}T$,

(here using the definition of $\beta(cT, bT, m)$). Thus,

$(c^{\beta(cT,bT,m)+d+r}R : b^m R) \subseteq c^{d+r}T \cap R = c^r(c^d T \cap R) \subseteq c^r R$.

Now suppose that we also have $T \subseteq R_b$. Pick $e \geq 1$ large enough that $b^e T \subseteq R$. Let $h = \beta(cT, bT, e)$, and $f = h + d$. By (6.19), it will suffice to show $\beta(cT, bT, m) \leq \beta(cR, bR, m) + f$. Since (6.4) shows that $\beta(cT, bT, m)$ is nondecreasing, it will suffice to show the stronger statement $\beta(cT, bT, m + e) \leq \beta(cR, bR, m) + f$. For this, by the minimality of $\beta(cT, bT, m + e)$ it will suffice to show that

$(c^{\beta(cR,bR,m)+f+r}T : b^{m+e}T) \subseteq c^r T$, for all $r \geq 1$. Pick

$x \in (c^{\beta(cR,bR,m)+f+r}T : b^{m+e}T)$. Since $b^e T \subseteq R$, $xb^e \in R$. Thus we have $xb^{m+e} \in c^{\beta(cR,bR,m)+f+r}T \cap R = c^{\beta(cR,bR,m)+h+r}(c^d T \cap R) \subseteq c^{\beta(cR,bR,m)+h+r}R$, (using the definitions of f and d). Therefore, we have $xb^e \in (c^{\beta(cR,bR,m)+h+r}R : b^m R) \subseteq c^{h+r}R$, (by the definition of $\beta(cR, bR, m)$). Thus $xb^e \in c^{h+r}T$. Since $h = \beta(cT, bT, e)$, we get $x \in (c^{h+r}T : b^e T) \subseteq c^r T$, as desired. The final conclusion is easy.

(6.22) Lemma. Let $b \in R$ be regular. Then $\alpha(bR, cR, m) = $ max $\{\alpha(bR_P, cR_P, m) \mid P \in$ Ass $R/bR\}$. In particular, $\lim \alpha(bR, cR, m)/m = $ max $\{\lim \alpha(bR_P, cR_P, m)/m \mid P \in$ Ass $R/bR\}$.

Proof. For $m \geq 1$, by definition of $\alpha(bR, cR, m)$, for any $k \geq 0$, $(b^m R : c^{\alpha(bR,cR,m)}R) = (b^m R : c^{\alpha(bR,cR,m)+k}R)$. For any prime ideal P of R, localizing at P shows that $(b^m R_P : c^{\alpha(bR,cR,m)}R_P) = (b^m R_P : c^{\alpha(bR,cR,m)+k}R_P)$. The minimality of $\alpha(bR_P, cR_P, m)$ therefore shows that $\alpha(bR_P, cR_P, m) \leq \alpha(bR, cR, m)$ for any prime P of R. If $h(m) = \max \{\alpha(bR_P, cR_P, m) \mid P \in \text{Ass } R/bR\}$, we have shown $h(m) \leq \alpha(bR, cR, m)$. For the other inequality, note that for all $P \in \text{Ass } R/bR$, and $k \geq 0$, we have $(b^m R_P : c^{h(m)}R_P) = (b^m R_P : c^{h(m)+k}R_P)$. It easily follows that $(b^m R : c^{h(m)}R) = (b^m R : c^{h(m)+k}R)$ (using that $\text{Ass } R/b^m R = \text{Ass } R/bR$, since b is regular). This shows that $\alpha(bR, cR, m) \leq h(m)$, proving the first conclusion. The second conclusion is an easy exercise.

We need some work of Samuel [S] and Nagata [N2].

(6.23) Remark. Let I and J be nonnilpotent ideals in a Noetherian ring R, with Rad I = Rad J. For $m \geq 1$, define $w(I, J, m)$ to be the smallest integer such that $J^{w(I, J, m)} \subseteq I^m$. Samuel shows that as $m \to \infty$, $\lim w(I, J, m)/m$ exists. Nagata shows that limit is rational. (Observe that since $J \subseteq \text{Rad } I$, the chain $(I^m : J^0) \subseteq (I^m : J^1) \subseteq (I^m : J^2) \subseteq \cdots$ eventually stabilizes to R, and the first term equaling R is $(I^m : J^{w(I, J, m)})$. Thus $w(I, J, m) = \alpha(I, J, m)$, so our work here is a generalization of that in [N2] and [S], in that we do not insist that $J \subseteq \text{Rad } I$.)

(6.24) Remark. In (8.4)(a), we prove that if $b \in R$ is regular, then there is a ring T with $R \subseteq T \subseteq R_b$, such that T is a finite R-module, and $E(bT) = \text{Ass } T/bT$.

(6.25) Proposition. Let b and c be regular elements of R. Suppose that $E(bR)$ consists of height 1 primes, (which holds if R is locally unmixed). Then as $m \to \infty$, $\lim \alpha(bR, cR, m)/m$ is rational.

Proof. By (6.24), there is a ring T with $R \subseteq T \subseteq R_b$ with T a finite R-module, and with $E(bT) = \text{Ass } T/bT$. By (1.13)(a), primes in $E(bT)$ lie over primes in $E(bR)$, and so $E(bT)$ consists of height 1 primes. Thus, by

(6.21), we may replace R by T. That is, we may assume that Ass R/bR = E(bR) = {P ∈ Spec R | b ∈ P and height P = 1}. By (6.22), we may assume that (R, P) is a local ring, with height P = 1. If c is not in P, then clearly α(bR, cR, m) = 0 for all m ≥ 1, and so lim α(bR, cR, m)/m = 0. On the other hand, if c is in P, then Rad bR = P = Rad cR. As is pointed out in

(6.23), in this case, α(bR, cR, m) = w(bR, cR, m), and as m → ∞, lim w(bR, cR, m)/m is rational.

(6.26) Corollary. Let b and c be regular elements of R. If the integral closure R' of R is a finite R-module with $R' \subseteq R_b$, then lim α(bR, cR, m)/m is rational.

Proof. By (6.21), lim α(bR, cR, m)/m = lim α(bR$'$, cR$'$, m)/m. Now, E(bR$'$) $\subseteq$ A*(bR$'$) = Ass R$'$/bR$'$. Since R$'$ is a Krull ring, primes in Ass R$'$/bR$'$ have height 1. Thus (6.25) implies that lim α(bR, cR, m)/m is rational.

(6.27) Proposition. Let R be locally unmixed, and let c ∈ R be regular. Then as m → ∞, lim β(I, cR, m)/m is rational.

Proof. By (6.14) and (6.19), lim β(I, cR, m)/m = lim α(cR(I), uR(I), m)/m. Since R is locally unmixed, so is R(I), [N1]. Thus (6.25) shows lim β(I, cR, m)/m is rational.

(6.28) Remark. Suppose there is an example where lim α(bR, cR, m)/m is not rational. By (6.24) and (6.21), we may assume that E(bR) = Ass R/bR. By (6.22), we may assume that (R, P) is local, with P ∈ Ass R/bR = E(bR). If (R*, P*) is the completion of R, then standard facts show that lim α(bR*, cR*, m)/m = lim α(bR, cR, m)/m. Since P ∈ E(bR), b is a maximal essential sequence in P. By (3.4), R* contains a depth 1 prime divisor of zero. Therefore, we may assume our hypothetical example is a complete local ring having a depth 1 prime divisor of zero.

The material in this chapter is taken from [M4] and [KM].

7 FINITE TRANSFORMS

Notation. Throughout this section, R will denote a Noetherian ring with total quotient ring $Q(R)$. We will consider a subset U of Spec R, and the transform determined by U, $T_U = \{u \in Q(R) \mid (R : u) \not\subseteq P$ for all $P \in U\}$, where $(R : u)$ consists of those elements $r \in R$ such that $ru \in R$. (If for each $P \in U$, $R - P$ consists of regular elements, so that $R_P \subseteq Q(R)$, then $T_U = \cap R_P$ over all $P \in U$.)

We now list several transforms of this sort which have drawn attention in the literature.

1) For I a regular ideal, let U consist of those primes not containing I. We easily see that equivalently, $T_U = \{u \in Q(R) \mid uI^n \subseteq R$ for some integer $n \geq 1\}$. Thus, $T_U = T(I)$ is the ideal transform of I used in chapter 5.

2) Let U be the set of height one primes. The resulting transform is denoted $R^{(1)}$.

3) Let U consist of all nonmaximal primes in Spec R. The resulting transform is denoted R^{ω}.

Recall that $\mathbf{P}(R)$ and $\mathbf{E}(R)$ were defined in chapter 5.

4) Let $U = \mathbf{P}(R)$. Here, $T_U = R$, as is well known. (We mention this example primarily to compare it to the next one, for the interplay between $\mathbf{P}(R)$ and $\mathbf{E}(R)$ is crucial to us.)

5) Let $U = \mathbf{E}(R)$. The resulting transform is denoted R^e.

(7.1) Lemma. $\mathbf{E}(R) \subseteq \mathbf{P}(R)$.

Proof. If $P \in \mathbf{E}(R)$, then by (5.1)(ii) and (iii), $P \notin \mathrm{Ass}\ R$ and $\mathrm{egr}\ P_P = 1$. By (3.1)(c), $\mathrm{gr}\ P_P = 1$, which implies $P \in \mathbf{P}(R)$.

The statement of theorem 7.2, and later results, is simplified by use of the next definition. Note that U and U^g determine the same transformation.

Definition. If $U \subseteq \mathrm{Spec}\ R$, let $U^g = \{P \in \mathrm{Spec}\ R \mid \text{there is a } P' \in U \text{ with } P \subseteq P'\}$.

(7.2) Theorem. Let $U \subseteq \mathrm{Spec}\ R$. The following are equivalent.

a) T_U is a finite R-module.

b) There is a finite subset $\{p_1, ..., p_n\}$ of $\mathbf{P}(R) - (\mathbf{E}(R)^g \cup U^g)$, consisting of regular primes, such that T_U equals the ideal transform

$T(p_1 \cap ... \cap p_n)$.

c) There are no regular primes in $\mathbf{E}(R) - U^g$, and only finitely many regular primes in $\mathbf{P}(R) - U^g$.

Proof. Let $T = T_U$.

a) $\Rightarrow$ b) Let $x_1, ..., x_m$ be module generators for T, and let

$I = (R : x_1) \cap ... \cap (R : x_m)$, which is a regular ideal since each x_j is in

$Q(R)$. Since $IT \subseteq R$, we have $T \subseteq T(I)$. Let $u \in T(I)$. Then

$I \subseteq \mathrm{Rad}\ (R : u)$. Since for each i, $(R : x_i) \not\subseteq P$ for all $P \in U$, we have

$I \not\subseteq P$, and so $(R : u) \not\subseteq P$. This shows that $u \in T$. Therefore, $T = T(I)$. If we now take $\{p_1, ..., p_n\}$ to be the primes minimal over I, obviously

$T = T(I) = T(p_1 \cap \ldots \cap p_n)$. Furthermore, any such p_i is not in U^g since I is not contained in any prime in U. Also, since $T = T(I)$ is a finite R-module, (5.2)(2) shows that I is not contained in any prime in $E(R)$, and so each p_i is not in $E(R)^g$. Finally, since p_i is minimal over $(R : x_j)$ for some $j = 1, \ldots, m$, we see that $p_i \in P(R)$, proving (b).

b) $\Rightarrow$ a) This is immediate from (5.2)(2).

a) $\Rightarrow$ c) Assume that (a), and hence (b), holds. Then T is Noetherian and equals $T(J)$, with $J = p_1 \cap \ldots \cap p_n$ as in (b). Let p be a regular prime in $E(R)$. We need $p \in U^g$. By (7.1), $p = (e : f)$ for some $e, f \in R$ with e regular. Let $S = R - p$. Since $p \in E(R)$, $J \not\subseteq p$, by (b). We easily see $T_S = T(J)_S = R_p$. We now see that $(e : f)R_p = p_p = q_S$ for some prime q of T having the form $q = (e :_T h)$, with $q \cap R = p$. As $(e :_T h) \neq T$, $h/e \notin T$. Therefore, for some $P \in U$, $(R : h/e) \subseteq P$. We claim that $(e :_T h) \cap (R - P) = \emptyset$. Otherwise, for some $s \in R - P$ and $t \in T$, $sh = et$. As $t \in T$, $(R : t) \not\subseteq P$, so there is an $s' \in R - P$ with $s'sh = e(s't) \in eR$. Thus $s's \in (R : h/e)$. This contradicts that $(R : h/e) \subseteq P$. Therefore $q = (e :_T h)$ is disjoint from $R - P$, as claimed. As $q \cap R = p$, we see that $p \subseteq P \in U$. That is, $p \in U^g$, and we have proved the first statement of (c). Next, suppose that $V = \{q \in P(R) - U^g \mid q \text{ is regular}\}$. We must show that V is finite. Each $q \in V$ has the form $q = (g : f)$ for some $f, g \in R$ with g regular. Since this ideal is not contained in P for all $P \in U$, we have $f/g \in T = T(J)$. Therefore $J \subseteq \text{Rad}(R : f/g) = q$. That is, every $q \in V$ contains J. If c is a regular element of J, then every $q \in V$ is a prime divisor of cR (since $q \in P(R)$ implies that q_q has grade 1). Therefore, V is finite.

c) $\Rightarrow$ b) Suppose (c) holds, and let $p_1, \ldots, p_n$ be all of the regular primes in $P(R) - U^g$. Since each p_i is not in U^g, (c) shows that each p_i is also not in $E(R)^g$. To prove (b), we will show that $T = T(p_1 \cap \ldots \cap p_n)$. First, suppose that v is in this ideal transform. Then $p_1 \cap \ldots \cap p_n \subseteq$

Rad $(R : v)$. Since each $p_j \notin U^g$, clearly $(R : v) \nsubseteq P$ for all $P \in U$.

Therefore, $v \in T$. Thus T contains our ideal transform. Next, pick

$w \in T$. We have that $(R : w) \nsubseteq P$ for all $P \in U$, and so if p is minimal

over $(R : w)$, then $p \notin U^g$. However, p is in $\mathbf{P}(R)$. Clearly p is regular,

and so equals one of $p_1, ..., p_n$. As this holds for all p minimal over

$(R : w)$, we see that $p_1 \cap ... \cap p_n \subseteq$ Rad $(R : w)$. Thus $w \in$

$T(p_1 \cap ... \cap p_n)$. Therefore this ideal transform equals T, so (b) holds.

Notation. Let $\mathbf{N}(R) = \mathbf{P}(R) - \mathbf{E}(R)$, and $\mathbf{M}(R) = \mathbf{P}(R) - \mathbf{E}(R)^g$.

(7.3) Corollary.   Let R be a Noetherian ring. Then

a) $R^{(0)}$ is a finite R-module if and only if there are only finitely many
maximal ideals in $\mathbf{P}(R)$, and none in $\mathbf{E}(R)$.

b) $R^{(1)}$ is a finite R-module if and only if every regular prime in $\mathbf{E}(R)$
has height 1, and only finitely many regular primes in $\mathbf{P}(R)$ have
height > 1.

c) R^e is a finite R-module if and only if $\mathbf{M}(R)$ is finite.

d) For any regular ideal I, $T(I) \cap R^e$ is a finite R-module.

Proof. a) This is immediate from (7.2) and the fact that any nonregular
maximal ideal is in Ass R, and so is not in $\mathbf{P}(R)$.

b) This is immediate from (7.2) and the fact that if a height 1 prime
contains a regular prime, the two are equal.

c) (7.2) immediately shows that R^e is a finite R-module if and only if
$\mathbf{M}(R)$ contains only finitely many regular primes. We now claim that
in general, $\mathbf{M}(R)$ can contain at most finitely many nonregular

primes. Otherwise, there would be a $Q \in$ Ass R which contains
infinitely many primes in $\mathbf{N}(R)$. This would violate (7.7), proved
below.

d) Clearly $T(I) \cap R^e = T_U$ with $U = \{P \in \text{Spec } R \mid I \not\subseteq P\} \cup E(R)$. We claim U satisfies condition (c) of (7.2), the part concerning $E(R)$ being trivial. Thus let P be a regular prime in $\mathbf{P}(R)$ which is not in U^g. Then $P \notin U$, and so $I \subseteq P$. If c is a regular element in I, then P is a prime divisor of cR (since P_P has grade 1). Therefore, such P are finite in number.

Remark. Concerning (7.3)(c), it always holds that R^e is the union of all $T(I)$ such that I is a regular ideal and $T(I)$ is a finite R-module, by [M2 , (10.21)].

In our next corollary, we take R to be a Noetherian ring with $\mathbf{N}(R)$ finite. (In (7.5) and (7.6) we will see that this condition is often satisfied.)

(7.4) Corollary. Let R be a Noetherian ring with $\mathbf{N}(R)$ finite. Then

a) If $U \subseteq \text{Spec } R$, T_U is a finite R-module if and only if there are no regular primes in $\mathbf{E}(R) - U^g$.

b) R^e is a finite R-module.

c) $R^{(\omega)}$ is a finite R-module if and only if no maximal ideal of R is in $\mathbf{E}(R)$.

d) $R^{(1)}$ is a finite R-module if and only if every regular prime in $\mathbf{E}(R)$ has height 1.

e) Suppose also that R is locally unmixed. Then $R^{(1)}$ is a finite R-module.

f) Suppose that R is a locally unmixed domain, and that the domain A is finitely generated over R. Then $A^{(1)}$ is a finite A-module.

Proof. Parts (a) through (d) follow easily from (7.2) and (7.3). For (e), if $P \in \mathbf{E}(R)$, by (5.1)(iii) R_P^* has a depth 1 prime divisor of zero. R_P being unmixed means that all primes in $\text{Ass } R_P^*$ have depth equal to height P. Thus, height P = 1. Therefore (e) follows from (d). Finally, for (f), the hypothesis and [N1, Corollary, p. 61] imply that A is locally unmixed. Also, (7.8) shows that $\mathbf{N}(A)$ is finite. Thus (e) implies (f).

(7.5) Lemma. Let R be a Noetherian domain whose integral closure R' is a finite R-module. Then $N(R)$ and $M(R)$ are finite.

Proof. Pick $0 \neq x \in R$ with $xR' \subseteq R$. Consider any $P \in P(R)$. If $x \notin P$, then because $R \subseteq R' \subseteq R_P$, we easily see that $P_P \cap R' \in P(R')$. Thus height $P_P \cap R' = 1$. It follows that height $P = 1$, so that $P \in E(R)$. This shows that if $P \in N(R)$, then $x \in P$. Each such P is a prime divisor of x, so $N(R)$ is finite. As $M(R) \subseteq N(R)$, $M(R)$ is also finite.

We come to the heart of this chapter.

(7.6) Theorem. Let $R \subseteq A \subseteq B$ be Noetherian rings with B finitely generated over R.

a) If R is semi-local, and if Ass $R = \{Q \cap R \mid Q \in$ Ass $A\}$, then $N(A)$ and $M(A)$ are finite.

b) If $N(A)$ is finite, then $N(R)$ is finite. The converse holds if Ass $R = \{Q \cap R \mid Q \in$ Ass $A\}$.

c) If Ass A has no embedded primes and $M(A)$ is finite, then $M(R)$ is finite.

d) Suppose Ass R has no embedded primes and Ass $R = \{Q \cap R \mid Q \in$ Ass $A\}$. Then $M(R)$ is finite if and only if $M(A)$ is finite.

The proof of (7.6) will be accomplished by proving the following five propositions.

(7.7) Proposition. Each member of Spec R contains only finitely many members of $N(R)$.

(7.8) Proposition. Let R be a domain and let the domain B be finitely generated over R. Then $N(R)$ (respectively $M(R)$) is finite if and only if $N(B)$ (respectively $M(B)$) is finite.

(7.9) Proposition. $N(R)$ is finite if and only if $N(R/Q)$ is finite for all $Q \in$ Ass R.

(7.10) Proposition. If $M(R)$ is infinite, then $M(R/Q)$ is infinite for some $Q \in$ Ass R.

(7.11) Proposition. If $M(R)$ is finite, then $M(R/Q)$ is finite for every Q which is maximal in Ass R.

Proof of (7.6). The arguments for the various parts are all quite similar. We content ourselves with proving one direction of (d). Specifically, assume that Ass R has no embedded primes and is the restriction to R of Ass A. Assume also that $M(A)$ is finite. We will show that $M(R)$ is finite. If not, then by (7.10) we have $M(R/Q)$ infinite for some $Q \in$ Ass R. Now it is always true that there is a $q_1 \in$ Ass A with $q_1 \cap R = Q$. Let q_2 be a maximal member of Ass A with $q_1 \subseteq q_2$. By hypothesis, $q_2 \cap R \in$ Ass R, and as Ass R has no embedded primes, we see that $q_2 \cap R$ must equal $q_1 \cap R = Q$. We can now find a prime $q \in$ Ass B with $q \cap A = q_2$. By (7.8) applied to $R/Q \subseteq B/q$, we have $M(B/q)$ infinite. Now that same proposition applied to $A/q_2 \subseteq B/q$ shows that $M(A/q_2)$ is infinite. As q_2 is maximal in Ass A, (7.11) shows that $M(A)$ is infinite. This is a contradiction. The remaining arguments for (7.6) are left to the reader. (Use (b) and (7.7) to prove (a).)

We now begin proving Propositions (7.7) through (7.11). We need two tools. The first is an interesting result about grade. The next lemma is a translation of part of [G, IV, 6.10.6] from the language of preschemes to that of commutative rings.

(7.12) Lemma. Let $Q \in$ Spec R. Then there is a $b \in R - Q$ such that for all $P \in$ Spec R with $Q \subseteq P$ and $b \notin P$, we have
$$\text{gr } P_P = \text{gr } Q_Q + \text{gr}(P_P/Q_P).$$

Recall that the concept of a conforming pair was defined in chapter 3. It will be our second tool. We give three lemmas involving conforming pairs.

(7.13) Lemma. ([M2, lemma 9.6]) Let I be an ideal of R and let V be an infinite set of primes of R, each of which contains I. Then there is a conforming pair (Q, W) with $I \subseteq Q$ and $W \subseteq V$.

Proof. By the ascending chain condition, expand I to an ideal Q maximal with respect to having $W = \{P \in V \mid Q \subset P\}$ infinite. Thus, if $W' \subseteq W$ with W' infinite, then $\cap \{P \in W'\} = Q$. It is easily seen that Q is prime, and (Q, W) is a conforming pair.

(7.14) Lemma. Let W be an infinite subset of Spec R and let $Q \in$ Spec R with $Q \subset P$ for all $P \in W$. Suppose that $\{P/Q \mid P \in W\}$ is a subset of $\mathbf{P}(R/Q)$. Then (Q, W) is a conforming pair.

Proof. Let W' be an infinite subset of W, and suppose that $\cap \{P \in W'\} \neq Q$. Pick x in that intersection but not in Q. For $P \in W'$ we have $x + Q \in P/Q \in \mathbf{P}(R/Q)$. Thus P/Q is a prime divisor of $x + Q$ for the infinitely many $P \in W'$. This is impossible.

(7.15) Lemma. Let (Q, W) be a conforming pair in R. The following are equivalent.

a) $P \in \mathbf{P}(R)$ for all but finitely many $P \in$ W.

b) $Q \in$ Ass R, and $P/Q \in \mathbf{P}(R/Q)$ for all but finitely many $P \in$ W.

Proof. Suppose (a) holds. Pick $b \notin Q$ as in (7.12). By (a) and the fact that (Q, W) is a conforming pair, we see that for all but finitely many $P \in$ W we have $b \notin P \in \mathbf{P}(R)$. Therefore, $1 = \mathrm{gr}\ P_P = \mathrm{gr}\ Q_Q + \mathrm{gr}\ (P_P/Q_P)$. Since $Q \neq P$, clearly $\mathrm{gr}\ Q_Q = 0$ and $\mathrm{gr}\ (P_P/Q_P) = 1$. Thus $Q \in$ Ass R, and $P/Q \in \mathbf{P}(R/Q)$. This proves (b). The converse is proved similarly.

(7.16) Lemma. Let R be a Noetherian ring, and let P be a prime with $P \notin$ Ass R. Then $P \in \mathbf{E}(R)$ if and only if $P/Q \in \mathbf{E}(R/Q)$ for some $Q \in$ Ass R with $Q \subset P$.

Proof. This follows from (5.1)(iii) and standard facts about completions.

Proof of (7.9). We will work with the contrapositive statements. Suppose $\mathbf{N}(R)$ is infinite. Now (7.13) (taking I = 0) gives a conforming pair (Q, W) with $W \subseteq \mathbf{N}(R)$. By (7.15), $Q \in$ Ass R and $P/Q \in \mathbf{P}(R/Q)$ for infinitely many $P \in W$. Since $P \in \mathbf{N}(R) \subseteq \mathbf{P}(R)$, $P \notin \mathbf{E}(R) \cup$ Ass R, and so by (7.16), we have $P/Q \notin \mathbf{E}(R/Q)$. This shows that $P/Q \in \mathbf{N}(R/Q)$ for infinitely many $P \in W$.

Conversely, suppose that $Q \in$ Ass R and that $\mathbf{N}(R/Q)$ is infinite. Let $W = \{P \in$ Spec R $\mid Q \subset P$ and $P/Q \in \mathbf{N}(R/Q)\}$. By (7.14), (Q, W) is a conforming pair, and by (7.15), $W' = \{P \in W \mid P \in \mathbf{P}(R)\}$ is an infinite subset of W. Now let $W'' = \{P \in W' \mid P \in \mathbf{E}(R)\}$. If we can show that W'' is finite, then $W' - W''$ will be an infinite subset of $\mathbf{N}(R)$, and we will be done. Therefore, suppose that W'' is infinite. Using (7.16), we easily see there is a $q \in$ Ass R and an infinite subset V of W'' of with $q \subset P$ and $P/q \in \mathbf{E}(R/q)$ for all $P \in V$. By (7.14), (q, V) is a conforming pair. Since $V \subseteq W$, (Q, V) is also a conforming pair. Thus, q = Q. If $P \in V$, we see that $P/Q = P/q \in \mathbf{E}(R/q) = \mathbf{E}(R/Q)$, contradicting that $P/Q \in \mathbf{N}(R/Q)$.

(7.17) Lemma. Let $R \subseteq B$ be a faithfully flat extension of Noetherian rings. Let $P \in$ Spec B and let $p = P \cap R$.

i) If $p \in \mathbf{E}(R)$, and P is minimal over pB, then $P \in \mathbf{E}(B)$.

ii) If $p \notin$ Ass R and $P \in \mathbf{E}(B)$, then $p \in \mathbf{E}(R)$.

Proof. Say $p \in \mathbf{E}(R)$ and P is minimal over pB. By definition, there is a $b \in R$ with $p \in \mathbf{E}(bR)$ and the image of b regular in R_p. By (1.9)(b), $P \in \mathbf{E}(bB)$, and since the image of b is regular in Bp, we have $P \in \mathbf{E}(B)$. This proves (i). For (ii), suppose $P \in \mathbf{E}(B)$. As $p \notin$ Ass R, pick $b \in p$ with the image of b regular in R_p, and hence also in Bp. By (5.1)(ii), $Pp \in \mathbf{E}(bBp)$, so that $P \in \mathbf{E}(bB)$. By (1.9)(a), $p \in \mathbf{E}(bR)$, and so $p \in \mathbf{E}(R)$.

Proof of (7.7). Suppose there were a counterexample. Localizing at the offending prime would produce a local ring R with $\mathbf{N}(R)$ infinite. Using (7.17)(ii), we see that we may take R to be complete. Using (7.9), we may further assume that R is a domain. Now it is well known that the integral closure of a complete local domain R is a finite R-module. Thus, (7.5) shows that $\mathbf{N}(R)$ is finite, a contradiction.

Proof of (7.10). If $\mathbf{M}(R)$ is infinite, then by (7.13) and (7.15), there is a $Q \in$ Ass R and an infinite subset W of $\mathbf{M}(R)$ such that for all $P \in W$, $Q \subset P$ and $P/Q \in \mathbf{P}(R/Q)$. Now (7.7) shows that at most finitely many members of W are contained in some prime in Ass R, and deleting these, we may assume that W consists of regular primes. For $P \in W$, we claim that $P/Q \in \mathbf{M}(R/Q)$. We need to show that $P/Q \notin \mathbf{E}(R/Q)^g$. If $P/Q \in \mathbf{E}(R/Q)^g$, then for some $P'/Q \in \mathbf{E}(R/Q)$, we have $P/Q \subseteq P'/Q$. By (7.16), we have $P' \in \mathbf{E}(R)$, so that $P \in \mathbf{E}(R)^g$. This contradicts that $P \in W \subseteq \mathbf{M}(R)$, and proves the claim. As W is infinite, we see now see that $\mathbf{M}(R/Q)$ is infinite.

Proof of (7.11). Suppose that $\mathbf{M}(R)$ is finite, but that $\mathbf{M}(R/Q)$ is infinite for some maximal member Q of Ass R. By (7.14), (7.15), and the fact that $\mathbf{M}(R)$ is finite, there is an infinite subset W of $\mathbf{P}(R) - \mathbf{M}(R)$ such that for all $P \in W$, $Q \subset P$ and $P/Q \in \mathbf{M}(R/Q)$. Let W' be the set consisting of primes in $\mathbf{E}(R)$ which contain at least one prime in W. As $W \subseteq \mathbf{P}(R)$, $W = (W \cap \mathbf{E}(R)) \cup (W \cap (\mathbf{N}(R)) = (W \cap W') \cup (W \cap \mathbf{N}(R))$. Thus, either $W \cap W'$ or $W \cap \mathbf{N}(R)$ is infinite. In the first case, it is clear that

W' is infinite. We now claim that W' is also infinite in the second case. To see this, suppose that $W \cap N(R)$ is infinite, and let P be in that intersection. As $P \in W \subseteq P(R) - M(R)$, we have $P \in E(R)^g$. Thus $P \subseteq P'$ for some $P' \in E(R)$. Of course $P' \in W'$. Now by (7.7), P' contains at most finitely many members of $N(R)$. Since $W \cap N(R)$ is infinite, the argument just given gives rise to infinitely many $P' \in W'$, proving the claim. By (7.16), there is a $q \in$ Ass R and an infinite subset V of W' such that for all $p \in V$, $q \subset p$ and $p/q \in E(R/q)$. By (7.14), (q, V) is a conforming pair. Now each prime in W' contains a prime in W, and so contains Q. As $V \subseteq W', Q \subseteq \cap \{p \in V\} = q$. As Q is maximal in Ass R, $Q = q$. For any $p \in V$ we now have $p/Q \in E(R/Q)$. However, $p \in V \subseteq W'$ implies there is a $P \in W$ with $P \subseteq p$. Since $P/Q \subseteq p/Q \in E(R/Q)$, we have contradicted that $P/Q \in M(R/Q)$.

(7.18) Lemma. Let R be a Noetherian domain and let X be an indeterminate over R. Then $N(R[X]) = \{PR[X] \mid P \in N(R)\}$ and $M(R[X]) = \{PR[X] \mid P \in M(R)\}$.

Proof. Since $R \subseteq R[X]$ is faithfully flat, using (7.17)(ii) it is easily seen that if $P \in N(R)$ then $PR[X] \in N(R[X])$. Let $q \in N(R[X])$, and let $p = q \cap R$. Then $p \neq 0$, since otherwise either $q = 0$, implying $q \notin P(R[X])$, or height $q = 1$, implying $q \in E(R[X])$. For $0 \neq b \in p, q \in P(R[X])$ implies that q is a prime divisor of $bR[X]$. It is well known that this implies $q = pR[X]$ and p is a prime divisor of bR, so that $p \in P(R)$. That $p \in N(R)$ now follows from (7.17)(i). This proves the first statement. The other is done similarly.

(7.19) Lemma. Let $R \subseteq B$ be a finite free integral extension of Noetherian domains. Then $N(B) = \{Q \in \text{Spec } B \mid Q \cap R \in N(R)\}$ and $M(B) = \{Q \in \text{Spec } B \mid Q \cap R \in M(R)\}$.

Proof. Let $0 \neq P \in$ Spec R, and let Q be any prime in Spec B lying over P. As our extension is integral, Q is minimal over PB. As our extension is faithfully flat, we see that $P \in \mathbf{P}(R)$ if and only if $Q \in \mathbf{P}(B)$. By (7.17), $P \in \mathbf{E}(R)$ if and only if $Q \in \mathbf{E}(B)$. The result is now easy, the second part using going up.

(7.20) Lemma. Let $0 \neq b$ be an element in a Noetherian domain R. Then $\mathbf{N}(R)$ (respectively, $\mathbf{M}(R)$) is finite if and only if $\mathbf{N}(R[1/b])$ (respectively, $\mathbf{M}(R[1/b])$) is finite.

Proof. Both $\mathbf{P}(R)$ and $\mathbf{E}(R)$ behave well with respect to localization. Also, if $b \in P \in \mathbf{P}(R)$, then $P \in$ Ass R/bR. Thus, only finitely many primes are lost in passing from $\mathbf{P}(R)$ to $\mathbf{P}(R[1/b])$. With these facts in mind, the proof is straightforward, except for one case, which we do now. Thus, suppose that $\mathbf{M}(R)$ is finite. We will show that $\mathbf{M}(R[1/b])$ is finite. Let W be the set of primes in Spec R which are not in $\mathbf{M}(R)$, but are the contractions of primes in $\mathbf{M}(R[1/b])$. Since a prime in $\mathbf{M}(R[1/b])$ contracts to either a prime in $\mathbf{M}(R)$ or a prime in W, and since $\mathbf{M}(R)$ is finite, it will suffice to show that W is finite. Note that $W \subseteq \mathbf{P}(R)$. Let W' consist of the primes in $\mathbf{E}(R)$ which contain at least one prime in W. Notice that each prime in W' must contain b, since that prime vanishes when passing to R[1/b]. Each prime in W' is a prime divisor of bR, and so W' is finite. $W = (W \cap \mathbf{E}(R)) \cup (W \cap \mathbf{N}(R)) = (W \cap W') \cup (W \cap \mathbf{N}(R))$. Clearly $W \cap W'$ is finite, and so it only remains to show that $W \cap \mathbf{N}(R)$ is finite. If $P \in W \subseteq \mathbf{P}(R) - \mathbf{M}(R)$, P is contained in some prime $P' \in \mathbf{E}(R)$. We see that $P' \in W'$. That is, each prime in $W \cap \mathbf{N}(R)$ is contained in some prime in W'. Since W' is finite, (7.7) shows $W \cap \mathbf{N}(R)$ is finite. Thus W is finite.

<u>Proof of (7.8)</u>. We may assume that $B = R[x]$. If x is transcendental over R, the result follows from (7.18). If x is algebraic over R, let b be the leading coefficient of a nonzero polynomial over R of minimal degree having x as a root. By (7.20), we may replace $R \subseteq B$ by $R[1/b] \subseteq B[1/b]$. This last is a finite free integral extension. (7.19) completes the proof.

We ask two (equivalent) questions.

Question 1. If $\mathbf{M}(R)$ is finite, will $\mathbf{M}(R/Q)$ be finite for all $Q \in \text{Ass } R$ (instead of just the maximal Q in Ass R as in (7.11))?

Question 2. Is there a strong version of (7.6)(d), in which the hypothesis that R have no embedded prime divisors of zero is deleted?

We suspect that the answer to both of these questions is no. We now show that the two questions are equivalent.

(7.21) Proposition. The answer to question 1 is yes if and only if the answer to question 2 is yes.

Proof. If the answer to question 1 is yes, then straightforward arguments using (7.8) and (7.10) show that the answer to question 2 is yes. Conversely, suppose that the answer to question 1 is no. Then for some R and $Q \in \text{Ass } R$, we have $\mathbf{M}(R)$ finite but $\mathbf{M}(R/Q)$ infinite.

Let $q_1 \cap \ldots \cap q_n = 0$ be a primary decomposition of zero, with Rad $q_1 = Q$. Let $A = R/q_1 \oplus \cdots \oplus R/q_n$, and embed R in A in the obvious way. Now A is a finite R-module, and primes in Ass A contract to primes in Ass R. Thus, $R \subseteq A$ satisfies the hypothesis of the "strong version" of (7.6)(d) which we are considering. We claim that it does not satisfy the conclusion, which will show that the answer to question 2 is no, as desired. By assumption, we have $\mathbf{M}(R)$ finite. We will show that $\mathbf{M}(A)$ is infinite. Were it finite, then since A has no embedded prime divisors of zero, by (7.11) we would have $\mathbf{M}(A/P)$ finite for all $P \in \text{Ass } A$. However, for $P = Q/q_1 \oplus R/q_2 \oplus \cdots \oplus R/q_n$, we have $A/P = R/Q$, and we have contradicted that $\mathbf{M}(R/Q)$ is infinite.

(7.22) Example. (R. Heitmann) We give an example of a Noetherian domain R, with $\mathbf{N}(R) = \mathbf{M}(R)$ infinite. In particular, by (7.3)(c), R^e is not a finite R-module. With K a field and with two sets of indeterminates $X = \{X_i \mid i = 1, 2, \cdots\}$ and $Y = \{Y_i \mid i = 1, 2, \cdots\}$, consider K[X, Y], and let D consist of those polynomials f such that if some X_i is a factor of some monomial of f, then either X_i^2 or Y_i is a factor of that monomial. Let $P_i = (X_i, Y_i)K[X, Y] \cap D$. Let S be the complement in D of the union of the P_j, and let $R = D_S$. We leave it to the reader to verify that the maximal ideals of R are exactly the P_jR, that R is Noetherian, and that each P_jR is in $\mathbf{P}(R)$. We claim that no P_jR is in $\mathbf{E}(R)$. The integral closure of R localized at P_jR is the finite extension $L[X_j, Y_j]$ localized at $(X_j, Y_j)L[X_j, Y_j]$, with $L = K(X - \{X_j\}, Y - \{Y_j\})$. The completion of this last ring obviously does not contain a depth 1 prime divisor of zero, and so our claim follows from (1.13)(a).

This material is taken from [MR2].

8 ESSENTIAL PRIMES AND PROJECTIVE EXTENSIONS

Definition. The ideals I and J are projectively equivalent if for some positive integers n and m, we have $I^n = J^m$. If for some positive integer n, we have $I^n \subseteq J \subseteq \overline{I^n}$, then we will call J a projective extension of I. (Note that a projective extension of I is projectively equivalent to I.)

The main goal of this chapter is to show there is a projective extension J of I with $A^*(J) = E(J) = E(I)$. This originally appeared in [KMOR].

(8.1) Lemma. Let I and J be projectively equivalent. Then $\overline{Q}^*(I) = \overline{Q}^*(J), \overline{A}^*(I) = \overline{A}^*(J), Q(I) = Q(J)$, and $E(I) = E(J)$.

Proof. $\overline{Q}^*(I) = \overline{Q}^*(J)$ and $Q(I) = Q(J)$ follow easily from the definitions and the fact that I and J have the same radical. In order to prove that $E(I) = E(J)$, it will be enough to show that for any ideal I, $E(I) = E(\bar{I})$, and for any $n \geq 1$, $E(I) = E(I^n)$.

Let $\mathbf{R} = \mathbf{R}(I)$ be the Rees ring of R with respect to I, and let $\mathbf{A} = R[u, \bar{I}t]$ be the Rees ring of R with respect to $\bar{I}$. Since I reduces $\bar{I}$, (6.5) shows that $\mathbf{A}$ is a finite $\mathbf{R}$-module. Since $\mathbf{A}$ is in the total quotient ring of $\mathbf{R}$, primes in Ass $\mathbf{A}$ contract to primes in Ass $\mathbf{R}$. By (1.13)(a), we have that $Q(u\mathbf{R}) = \{q \cap \mathbf{R} \mid q \in Q(u\mathbf{A})\}$. Now $R \subseteq \mathbf{R} \subseteq \mathbf{A}$. Let $P \in E(I)$. By definition, there is a $q' \in Q(u\mathbf{R})$ with $q' \cap R = P$. For some $q \in Q(u\mathbf{A})$, $q \cap \mathbf{R} = q'$. Since $q \cap R = P$, the definition of $E(\bar{I})$ shows that $P \in E(\bar{I})$. Thus $E(I) \subseteq E(\bar{I})$. The reverse inclusion is similar, showing $E(I) = E(\bar{I})$.

77

Let $B = R[u^n, I^n t^n]$. There is an obvious isomorphism between B and $R[u, I^n t]$, the Rees ring of R with respect to I^n. Using the definition of $E(I^n)$ and this isomorphism, we see that $E(I^n) = \{q' \cap R \mid q' \in Q(u^n B)\}$. Since $R = B[u, IT]$, and since $u^n \in B$ and $(It)^n \subseteq B$, we see that $B \subseteq R$ is a finite module extension. It is easily seen that primes in Ass R contract to primes in Ass B, since it holds for Ass $R[u, t]$ and Ass $R[u^n, t^n]$. By 1.13(a), $Q(u^n B) = \{q \cap B \mid q \in Q(u^n R)\}$ $= \{q \cap B \mid q \in Q(uR)\}$, the last equality since $Q(uR) = Q(u^n R)$ (as in the first sentence of this proof). We have $R \subseteq B \subseteq R$. If $P \in E(I^n)$, then by the preceding, there is a $q' \in Q(u^n B)$ with $q' \cap R = P$, and there is a $q \in Q(uR)$ with $q \cap B = q'$. As $q \cap R = P$, the definition of $E(I)$ shows that $P \in E(I)$. Therefore, $E(I^n) \subseteq E(I)$. The reverse containment is argued similarly. This shows $E(I) = E(I^n)$.

A proof that $\overline{A}^*(I) = \overline{A}^*(J)$ could be given similar to the above. Instead, we use (1.2). It easily shows that $\overline{A}^*(I) = \overline{A}^*(\overline{I})$ and $\overline{A}^*(I) = \overline{A}^*(I^n)$ for any ideal I and any $n \geq 1$, and our result follows.

If J is projectively equivalent to I, it may well happen that $A^*(J) \neq A^*(I)$. However, by (8.1) and (1.1)(d), we have $E(I) = E(J) \subseteq A^*(J)$. Thus we see that $E(I) \subseteq \cap A^*(J)$ where the intersection is over all J projectively equivalent to I. In [Kz1], D. Katz showed that equality holds; i.e., $E(I) = \cap A^*(J)$. Actually, he did more. When [Kz1] was written, $E(I)$ had not yet been defined. Katz studied $\cap A^*(J)$ over J projectively equivalent to I (actually over J which are projective extensions of I), and showed that the primes in that set were exactly those primes which satisfied the definition given here for $E(I)$. Thus credit for the explicit concept of $E(I)$ is due to him. (The concept was implicit in Schenzel's work in [Sc].) The result we will show here, that there is a projective extension J of I with $A^*(J) = E(I)$, is stronger than Katz's original result, and our arguments here are independent of those in [Kz1].

Definition. Suppose $C \subseteq D$ are rings and x is an element of C such that $C_x = D_x$. Let U be a subset of Spec C, and suppose that $x \notin P$ for all $P \in U$.

Let $V = \{P_x \cap D \mid P \in U\}$. Then we shall say that x lifts U to V. (In this case, there is a natural one-to-one inclusion preserving correspondence between U and V, corresponding primes having the same height).

(8.2) Lemma. Let C be a Noetherian ring and let b be a regular element of C. Let $E(bC) = \{Q_1, ..., Q_m\}$, and let $P_1, ..., P_n$ be the prime divisors of bC which are not contained in $Q_1 \cup ... \cup Q_m$. Let x be a regular element in $(P_1 \cap ... \cap P_n) - (Q_1 \cup ... \cup Q_m)$. (Such an x always exists.) Let $D = C_x \cap C_b$. Then

(i) D is a finite C-module.

(ii) $C_x = D_x$.

(iii) x lifts $\{P \in$ Ass C/bC $\mid$ P is contained in some prime in E(bC)$\}$ to Ass D/bD. Also, no prime in Ass D/bD contains x.

(iv) x lifts E(bC) to E(bD).

(v) the maximal members of Ass D/bD are identical to the maximal members of E(bD).

(vi) If C satisfies $\mathbf{R} = R(I) \subseteq C \subseteq R[u, t]$ with C graded, and if b = u, then x can be chosen to be homogeneous, and D will be a graded ring with $\mathbf{R} \subseteq C \subseteq D \subseteq R[u, t]$.

Proof. We first show that we can always find an x as in the statement. Since b is regular, we see that $P_1 \cap ... \cap P_n \nsubseteq$ $\cup\{Q \in$ Ass C$\} \cup (Q_1 \cup ... \cup Q_m)$, and so we use the prime avoidance lemma, (taking x = 1 if n = 0).

(i) One easily sees that D is the ideal transform T((b, x)C). If $q \in E(C)$, then by (5.1), q_q has essential grade 1. Thus, if $b \in q$, then $q \in E(bC)$, so that $x \notin q$. Therefore, no prime in $E(C)$ contains (b, x)C. Therefore, by (5.2), D is a finite C-module.

(ii) This is trivial.

(iii) Since b is a unit in C_b, $bD = b(C_x \cap C_b) = bC_x \cap C_b =$

$bC_x \cap C_x \cap C_b = bC_x \cap D = bD_x \cap D$. The second statement in (iii) follows easily from this. It is now clear that the prime divisors of bD are exactly the contractions of the prime divisors of $bD_x = bC_x$. Thus, we see that x lifts $\{P \in$ Ass $C/bC \mid x \notin P\}$ to Ass D/bD. The choice of x shows that $\{P \in$ Ass $C/bC \mid P$ is contained in some prime in $E(bC)\} =$ $\{P \in$ Ass $C/bC \mid x \notin P\}$, so that the first statement in (iii) is true.

(iv) By (1.1)(d), $E(bD) \subseteq A^*(bD) =$ Ass D/bD (since b is regular in D), and so primes in $E(bD)$ do not contain x, by (iii). Also, primes in $E(bC)$ do not contain x, by construction. Therefore, since $C_x = D_x$, it follows trivially from (1.1)(a) that x lifts $E(bC)$ to $E(bD)$.

(v) This follows easily from (iii), (iv), and the fact that $E(bD) \subseteq$ Ass D/bD.

(vi) Since u is homogeneous, the primes $Q_1, ..., Q_m$, are all homogeneous, as are the primes in Ass C. An easy variation of the standard prime avoidance lemma allows us to pick our x to be homogeneous. Obviously $C \subseteq D \subseteq C_u = R[u, t]$. Since $y \in D$ exactly when $y \in R[u, t]$ and some positive power of the homogeneous element x sends y into C, we easily see that D is a graded ring.

(8.3) Lemma. Let $A \subseteq B \subseteq C$ be Noetherian rings with $A \subseteq B$ a faithfully flat extension, and $B \subseteq C$ a finite module extension such that for all $z \in$ Ass C, $z \cap B \in$ Ass B. Suppose also that C is locally unmixed. Let b be a regular element of A (so b is still regular in C). Pick x and $D = C_x \cap C_b$ as in (8.2). Let $F = D \cap A_b$. Then F is a finite A-module contained in A_b, and Ass $F/bF = E(bF)$.

Proof. Since D is a finite C-module, it is a finite B-module. Thus B[F] is a finite B-module. We claim that B[F] is isomorphic to $B \otimes_A F$. The map from $B \times F$ to B[F] sending (b, f) to bf is balanced, and so induces a homomorphism from $B \otimes_A F$ to $B[F]$, sending $\Sigma b_i \otimes f_i$ to $\Sigma b_i f_i$. This homomorphism is easily seen to be onto. To show it is one-to-one, if $\Sigma b_i f_i = 0$, we need that $\Sigma b_i \otimes f_i = 0$. Since $F \subseteq A_b$, by flatness we have $B \otimes_A F \subseteq B \otimes_A A_b$. Thus, we may work in the latter ring. It is now an easy exercise that $\Sigma b_i f_i = 0$ implies $\Sigma b_i \otimes f_i = 0$ proving the claim. Since $B[F]$ is a finite B-module, faithful flatness now shows that F is a finite A-module.

Note that $F = C_x \cap A_b$. Since b is a unit in both C_b and A_b, $bD = bC_x \cap C_b$ and $bF = bC_x \cap A_b$. Thus it is easy to verify that $bD \cap F = bF$. Therefore, primes in Ass F/bF lift to primes in Ass D/bD.

If $Q \in E(bC)$, then Q_Q has essential grade 1. Since C_Q is unmixed, (3.7) shows that height $Q = 1$. Thus every prime in $E(bC)$ has height 1. By (8.2)(iv), all of the primes in $E(bD)$ have height 1. By (8.2)(v), we see that Ass $D/bD = E(bD)$.

Since $B \subseteq B[F] \subseteq D$, D is a finite $B[F]$-module. Also, primes in Ass D contract to primes in Ass B[F], since this holds between B and C, and $B \subseteq B[F] \subseteq B_b$ and $C \subseteq D \subseteq C_b$. By (1.13)(a), primes in $E(bD)$ contract to primes in $E(bB[F])$. Also, $B[F] \approx B \otimes_A F$, so by (1.9)(a), primes in $E(bB[F])$ contract to primes in $E(bF)$. Thus primes in $E(bD)$ contract to primes in $E(bF)$.

Combining the conclusions of the previous three paragraphs shows that Ass $F/bF = E(bF)$ (since one inclusion is by (1.1)(d)).

(8.4) Proposition. a) Let b be a regular element of the Noetherian ring R. Then there is a ring T with $R \subseteq T \subseteq R_b$ such that T is a finite R-module and Ass $T/bT = E(bT)$.

b) Let I be an ideal in the Noetherian ring R, and let **R** be the Rees ring of R with respect to I. There is a graded ring **T** with $\mathbf{R} \subseteq \mathbf{T} \subseteq R[u, t]$, such that **T** is a finite **R**-module, and Ass $\mathbf{T}/u\mathbf{T} = E(u\mathbf{T})$.

Proof. (a) Let $S = R - \cup\{p \in E(bR)\}$, and let $A = R_S$. As A is semi-local, let B equal the completion A^*. Let $q_1 \cap \ldots \cap q_n$ be a primary decomposition of 0 in B, and let $C = B/q_1 \oplus \ldots \oplus B/q_n$. There is a natural embedding of B into C. Under it, we see that b, A, B, and C satisfy the hypotheses of (8.3) (since every maximal localization of C is a complete local ring with a single prime divisor of zero, and hence is unmixed). Let F be as defined in that lemma. Then $R_S \subseteq F \subseteq (R_S)_b = (R_b)_S$. Also, if R' is the integral closure of R, then since F is a finite R_S-module, $F \subseteq R'_S$. Thus $R_S \subseteq F \subseteq (R_b \cap R')_S$. It is easy to find a ring G finitely generated over R with $R \subseteq G \subseteq R_b \cap R'$, such that $G_S = F$. Obviously, G is a finite R-module.

We now claim that $\{P \in \text{Ass } G/bG \mid P \text{ is contained in a prime in } E(bG)\} = E(bG)$. Suppose P is in the first set, and that $P \subseteq Q \in E(bG)$. By (1.13)(a) (and the fact that G is in the total quotient ring of R), $Q \cap R \in E(bR)$, and so is disjoint from S. Thus $P \cap S = \varnothing$. Therefore, P_S is a prime divisor of $bG_S = bF$. But Ass F/bF = E(bF), so $P_S \in E(bF) = E(bG_S)$, and so $P \in E(bG)$, by (1.1)(a). This shows one containment of our claim. The other is by Lemma 1.1(d), and the fact that $A^*(bG) = \text{Ass } G/bG$, since b is regular in G.

We now apply (8.2), starting with the ring G and $b \in G$. Pick x as in that lemma, and let $T = G_x \cap G_b$. By (8.2)(i), T is a finite G-module, hence a finite R-module. Clearly $T \subseteq G_b = R_b$. By the claim proved in the preceding paragraph and by (8.2)(iii), x lifts E(bG) to Ass T/bT. However, by (8.2)(iv), x lifts E(bG) to E(bT). Therefore, Ass T/bT = E(bT), as desired.

(b) Were we to simply apply part (a) to the ring **R** and the regular element u, we would find a ring **T** with $\mathbf{R} \subseteq \mathbf{T} \subseteq \mathbf{R}_u = R[u, t]$, such that **T** is a finite **R**-module, and Ass **T**/u**T** = E(u**T**). Thus **T** would have all the properties we want, except that of being graded. Therefore, this proof shall consist of an outline of what minor changes must be made in the proof of part (a) in order to assure that the resulting **T** is graded.

Let $S = R - \cup \{p \in E(I)\}$. Let $A = R_S$, $B = A^*$, and

$C = B/q_1 \oplus \ldots \oplus B/q_n$, where $q_1 \cap \ldots \cap q_n$ is a primary decomposition of 0 in B. Let $\mathbf{A} = A[u, IAt]$, $\mathbf{B} = B[u, IBt]$ and $\mathbf{C} = C[u, ICt]$. Now u, **A**, **B**, and **C** satisfy the hypotheses of (8.3), (since C being locally unmixed implies the same is true of **C**, [N1]). We apply (8.2)(vi) to u and **C**, and find a graded ring **D** with $\mathbf{C} \subseteq \mathbf{D} \subseteq C[u, t]$ as described in (8.2). We now let $\mathbf{F} = \mathbf{D} \cap \mathbf{A}_u$. By (8.3), Ass **F**/u**F** = E(u**F**), and **F** is a finite **A**-module with $F \subseteq A_u = A[u, t]$. Also, F is easily seen to be graded. Since $R_S = A$, we now find G with $G_S = F$, with $R \subseteq G \subseteq R[u, t]$, and with G a finite R-module, also insisting that G is graded. This last is easily done, since $S \subseteq R$. Primes in $E(uG)$ contract to primes in $E(uR)$, and then to primes in $E(I)$ (this last since $E(uR) = Q(uR)$ by (3.14)). Thus primes in $E(uG)$ are disjoint from S. As in the proof of part (a), we see that $E(uG) = \{P \in Ass\ G/uG\ |\ P$ is contained in some prime in $E(uG)\}$. As in part (a), apply (8.2)(vi) to G, to find the desired T.

We turn to the main goal of this chapter.

(8.5) Theorem. Let I be an ideal in a Noetherian ring R. Then there is a projective extension J of I such that if $R(J)$ is the Rees ring of R with respect to J, then $Ass\ R(J)/uR(J) = E(uR(J))$. Furthermore, $E(I) = A^*(J) = \cup Ass\ R/J^m$, over all $m \geq 1$.

Proof. Let $R \subseteq T \subseteq R[u, t]$ be as in (8.4)(b). Let $I_n = u^n T \cap R$. By (6.5), for large enough n and all $k \geq 1$, we have $I_n^k = I_{nk}$. For such a large n, let $J = I_n$. Since $R \subseteq T \subseteq R' \cap R[u, t]$, we see that $I^n \subseteq J \subseteq \overline{I^n}$, so that J is a projective extension of I. Let $B = R[u^n, Jt^n] \subseteq T$. Now it is easy to see that $u^n T \cap B = u^n B$. Thus primes in $Ass\ B/u^n B$ lift to primes in $Ass\ T/u^n T$.

Clearly $R[u^n, I^n t^n] \subseteq R = R[u, It]$ is a finite module extension. Since T is a finite R-module, it is a finite $R[u^n, I^n t^n]$-module. As $R[u^n, I^n t^n] \subseteq B \subseteq T$, we see that T is a finite B-module. As t is an indeterminate, we easily see that primes in $Ass\ T$ contract to primes in $Ass\ B$. By (1.13)(a), primes in $E(u^n T)$ contract to primes in $E(u^n B)$. Combining this fact with the conclusion of the preceding paragraph, and the fact that $Ass\ T/u^n T = Ass\ T/uT = E(uT) = E(u^n T)$ (the last equality by (8.1)), we see that primes in $Ass\ B/u^n B$ are in $E(u^n B)$, so that $Ass\ B/u^n B = E(u^n B)$. Now $B = R[u^n, Jt^n]$ is obviously isomorphic to $R(J) = R[u, Jt]$, the isomorphism taking u^n to u, and so the first conclusion of our result is true.

For the second conclusion, by (8.1) and (1.1)(d), we see that $E(I) = E(J) \subseteq A^*(J) \subseteq \cup Ass\ R/J^m$ over $m \geq 1$. Now let P be a prime divisor of J^m for some $m \geq 1$. As $J^m = u^m R(J) \cap R$, P lifts to a prime divisor Q of $u^m R(J)$. As u is regular, Q is a prime divisor of $uR(J)$. By the first conclusion, already proved, $Q \in E(uR(J)) = Q(uR(J))$, the equality by (3.14). By definition, $P = Q \cap R \in E(J) = E(I)$. Thus $\cup$Ass $R/J^m \subseteq E(I)$, which proves the second conclusion.

(8.6) Proposition. Let b be a regular element of R, and let P be a prime of R with $b \in P$. Let R' be the integral closure of R. Then

a) $P \in E(bR)$ if and only if P lifts to a prime in Ass T/bT for any ring T with $R \subseteq T \subseteq R'$ and with T a finite R-module.

b) $P \in \overline{A}^*(bR)$ if and only if P lifts to a prime in Ass T/bT for any ring T with $R \subseteq T \subseteq R'$. (Here, since T may not be Noetherian, we specify that we are using the Nagata definition of prime divisor.)

Proof. a) Suppose that $P \in E(bR)$. If $R \subseteq T \subseteq R'$, with T a finite R-module, then (1.13)(a) and (1.1)(d) show that P lifts to a prime $Q \in E(bT) \subseteq A^*(bT) = $ Ass T/bT, the equality since b is regular in T. Conversly, suppose that P is a prime in R, and that P lifts to a prime Q in Ass T/bT for any such T. Consider T as given in (8.4)(a). Since Ass T/bT = E(bT), (1.13)(a) shows that $P = Q \cap R \in E(bR)$.

b) The key to this is the following fact: $P \in \overline{A}^*(bR)$ if and only if there is a height 1 prime of R' lying over P. To see this, suppose that $P \in \overline{A}^*(bR)$. By (1.2), for large n, $P \in$ Ass R/ $b^n R$. As $b^n R' \cap R = b^n R$, and as R' is a Krull ring, we see that P lifts to a height 1 prime divisor of $b^n R'$. Conversely, if there is a height 1 prime of p of R' lying over P, then let w be in p, but in none of the other (finitely many) primes of R' which lie over P. It is easy to see that $p \cap R[w]$ has height 1. Obviously $p \cap R[w] \in \overline{A}^*(bR[w])$, and so by (1.13)(c), $P \in \overline{A}^*(bR)$.

Now let $P \in \overline{A}^*(bR)$ and suppose $R \subseteq T \subseteq R'$. By the
preceding, there is a height 1 prime p in R' with $p \cap R = P$. Let $Q = p \cap T$.
We will show that Q is a (Nagata) prime divisor of bT. (Remark: If T
were Noetherian, this would be trivial, since the previous paragraph
would show that in fact $Q \in \overline{A}^*(bT)$.) Only finitely many primes of R'
lie over P, and so this is also true of Q. Therefore, there is an
appropriate $w \in p$, such that if $D = T[w]$ and $q = p \cap D$, then height $q = 1$. It
does no harm to assume that R is local at P, so that P, p, Q, and q are all
maximal ideals. Only finitely many primes of D are minimal over bD
(since each such lifts to a prime minimal over bR'), and so we pick
$x \in D - q$ with x in every other prime of D minimal over bD. As D is a
finite T-module, there is a regular element $c \in T$ with $cD \subseteq T$. If z is a
minimal prime of D with $z \subseteq q$, then the intersection of all powers of q is
contained in z, (since D/z is contained in a discrete valuation ring
whose maximal ideal contracts to q/z). In particular, this shows that
there is an $m \geq 1$ such that $c \notin q^m$. Claim: for all $n \geq 1$, $cx^n \in T - b^m T$.
Since $cD \subseteq T$, certainly $cx^n \in T$. Suppose that $cx^n \in b^m T$. Then
$cx^n \in q^m$. However, $x^n \notin q$, and since q is maximal, q^m is
q-primary. This shows that $c \in q^m$, which is a contradiction, and so
proves the claim. Now let a be any element of Q. Since $a \in q$, xa is in
the nilradical of $b^m D$, so that for some $n \geq 1$, $(xa)^n \in b^m D$. Thus,
$cx^n a^n \in b^m cD \subseteq b^m T$. In view of the claim just proved, we see that a^n
is a zero divisor modulo $b^m T$. As b is regular in T, it follows that a^n,
and hence also a, are zero divisors modulo bT. Since this holds for all
$a \in Q$, and since Q is maximal, Q is a (Nagata) prime divisor of bT.
Conversely, suppose P lifts to a prime divisor of bT for all
$R \subseteq T \subseteq R'$. Letting $T = R'$, P lifts to a prime divisor of bR'. As R' is a
Krull ring, P lifts to a height 1 prime of R'. By the first paragraph,
$P \in \overline{A}^*(bR)$.

9 PERSISTENT PRIMES AND PROJECTIVE EXTENSIONS

Let W be a finite set of primes, and suppose that each prime in W contains the ideal I, here assuming that I is regular. In this chapter, we explore what conditions are needed on W to assure that there is a projective extension K of I with $A^*(K) = W$. The first such condition is trivial to find. By (8.1) and (1.1)(d), for any projective extension K of I, we have $E(I) \subseteq A^*(K)$. Thus, we are forced to assume that $E(I) \subseteq W$. We will write $W = U \cup E(I)$, with $U \cap E(I) = \varnothing$. We will show that under rather mild assumptions on U, there is a projective extension K of I with $A^*(K) = U \cup E(I)$. (If $U = \varnothing$, so that $W = E(I)$, then by (8.5) there is a projective extension K of I such that $A^*(K) = W$. Therefore, the results of this chapter can be viewed as a generalization of that theorem.) The results we are about to present first appeared in [MR3].

Notation. Throughout this section, I will be a regular ideal in a Noetherian ring R, and U will be finite set of prime ideals of R, each of which contains I. We will also assume that $U \cap E(I) = \varnothing$.

Definition. The ideal I is called prenormal if $I^n = \overline{I^n}$ for all large n. (Recall that I is called normal if $I^n = \overline{I^n}$ for all $n \geq 1$. In [M2], I is called prenormal if for some $m \geq 1$, I^m is normal. For regular ideals, the two definitions are equivalent, as is shown below.)

87

(9.1) Lemma.   Let I be regular.   Then

a) I is prenormal if and only if $\overline{I^n} = I^n$ for infinitely many n.

b) If I is not prenormal, then I_q is not prenormal for some $q \in A^*(I)$.

c) There is an ideal H such that for any prime P containing I, I_P is not prenormal if and only if $H \subseteq P$. (Here, $H = R$ if I is prenormal.)

d) Suppose that P is a prime ideal, and that I_P is prenormal. Then there is a projective extension K of I such that $P \in A^*(K)$ if and only if $P \in \overline{A}^*(I)$ if and only if $P \in E(I)$ if and only if $P \in A^*(I))$. In particular, if $P \in A^*(I) - \overline{A}^*(I)$, then I_P is not prenormal.

Proof. [M2, Proposition 11.15] shows that some power of I is normal if and only if $\overline{I^n} = I^n$ for all large n if and only if $\overline{I^n} = I^n$ for infinitely many n. This proves (a) (and also the equivalence of the two definitons of prenormal, when I is regular).   For (b), note that if I is not prenormal, then (a) shows that for any large n, $\overline{I^n} \neq I^n$. Standard arguments show there is a prime divisor q of I^n such that $\overline{I_q{}^n} \neq I_q{}^n$.

For n sufficiently large, we have $q \in A^*(I)$. As there are finitely many such q, one of them works for infinity many n. Part (a) now shows that I_q is not prenormal. For (c), let H be the product of all $q \in A^*(I)$ with I_q not prenormal. (H = R if there are no such q.) Let $I \subseteq P \in$ Spec R. If I_P is not prenormal, then (b) (applied in R_P) shows that $H \subseteq P$. Conversly, if $H \subseteq P$, then there is a $q \subseteq P$ with I_q not prenormal, and it easily follows that I_P is not prenormal. For the first part of (d), suppose that I_P is prenormal, and $P \in A^*(K)$ with K a projective extension of I. Suppose $I^h \subseteq K \subseteq \overline{I^h}$. For large n, $I_P{}^{hn} \subseteq K_P{}^n \subseteq \overline{I_P{}^{hn}} = I_P{}^{hn}$, the equality by prenormality. Thus, $K_P{}^n = \overline{I_P{}^{hn}}$. Since $P \in A^*(K)$, P_P is a prime divisor of $K_P{}^n = \overline{I_P{}^{hn}}$ (large n), so that $P \in \overline{A}^*(I) \subseteq E(I) \subseteq A^*(I)$. This shows one direction of all three equivalences. For the converses, simply note that if $P \in A^*(I)$, we may take K to be I. The second part of (d) follows immediately from the first.

After a convenient definition, we will state our main result.

Definition. An element in a partially ordered set is called an isolated element of that set if it is both minimal and maximal in the ordering.

(9.2) Theorem. Let I be regular, and let U_0 be the set of minimal members of U. Suppose that if $Q \in U_0$, then I_Q is not prenormal. Suppose also that if $Q \in U_0$ is not isolated in U, then for all large n, $\overline{Q_Q\, I_Q{}^n} \nsubseteq I_Q{}^n$. Then there is a projective extension K of I such that $A^*(K) = U \cup E(I)$.

The proof of (9.2) is a bit arduous, and so we defer it until the end of the chapter, first giving some applications. Our next remarks show that the first hypothesis of (9.2) is necessary, while something like the second hypothesis is also needed.

(9.3) Remarks: a) With notation as in (9.2), let $Q \in U_0$. Recall that our goal is to find a projective extension K of I, such that $A^*(K) = U \cup E(I)$. Suppose that such a K exists. Then $Q \in A^*(K)$. Were I_Q prenormal, then by (9.1)(d)), $Q \in E(I)$. This contradicts that $U \cap E(I) = \varnothing$. Therefore, if the conclusion of (9.2) is to hold, we must have that I_Q is not prenormal for all $Q \in U_0$.

b) Suppose now that $Q \in U_0$ is not isolated in U. As example (10.8) shows, it is not enough to merely assume that I_Q not prenormal. Something stronger is needed. We do not know the best assumption to make, but we suspect that the assumption made in (9.2), that for all large n, $\overline{Q_Q\, I_Q{}^n} \nsubseteq I_Q{}^n$, is fairly close to the best assumption. See part (c).

c) By (9.1)(a), I_Q not prenormal means for all large n, $\overline{I_Q{}^n} \nsubseteq I_Q{}^n$. The condition $\overline{Q_Q\, I_Q{}^n} \nsubseteq I_Q{}^n$ is a slight strengthening of that.

We now give some corollaries to (9.2). The situation when every element of U is isolated in U is particularly well behaved, and so we treat it first.

(9.4) Corollary. Let I be regular. Suppose that every member of U is isolated in U. Then the following are equivalent.

i) There is a projective extension K of I with $A^*(K) = U \cup E(I)$.

ii) There is a projective extension K of I with $U \subseteq A^*(K)$.

iii) For each $P \in U$, there is a projective extension K of I with $P \in A^*(K)$.

iv) For each $P \in U$, I_P is not prenormal.

Proof. (i) $\Rightarrow$ (ii) $\Rightarrow$ (iii) are obvious, and (iii) $\Rightarrow$ (iv) is by (9.1)(d) and the fact that $U \cap E(I) = \varnothing$. Finally, (iv) $\Rightarrow$ (i) follows trivially from (9.2) and the assumption that every prime in U is isolated in U.

(9.5) Remark. The implication (iii) $\Rightarrow$ (ii) of (9.4) does not hold without the assumption that every prime in U is isolated. An example appears in (10.8).

(9.6) Corollary. Let I be regular, and let U_0 be the set of minimal members of U. Suppose that for all $Q \in U_0$, I_Q is not prenormal. Suppose also that if $Q \in U_0$ is not isolated in U, then any one of the following conditions holds.

(a) $Q \notin A^*(I)$.

(b) There is a $q \in A^*(I) - \overline{A}^*(I)$ with $I \subseteq q \subset Q$.

(c) There is a prime q with $I \subseteq q \subset Q$ and I_q not prenormal.

Then there is a projective extension K of I such that $A^*(K) = U \cup E(I)$.

Proof. By (9.1)(d), we see that (b) $\Rightarrow$ (c). We also claim that (a) $\Rightarrow$ (c) (for $Q \in U_0$). Since $Q \in U_0$, the hypothesis shows that I_Q is not prenormal. By (9.1)(b), there is a $q_Q \in A^*(I_Q)$ with I_q not prenormal. Since $q \subseteq Q$ and $q \in A^*(I)$, while $Q \notin A^*(I)$, we have $q \subset Q$. Thus (c) holds. Therefore, we may assume that (c) holds for all $Q \in U_0$ with Q not isolated in U. We now claim that for such Q, $Q_Q \overline{I_Q{}^n} \not\subseteq I_Q{}^n$ for all large n. If not, then for infinitely many n, $Q_Q \overline{I_Q{}^n} \subseteq I_Q{}^n$. Let q be any prime with $I \subseteq q \subset Q$. Localizing our inclusion at q gives $\overline{I_q{}^n} = I_q{}^n$ (one inclusion being automatic) for infinitely many n. By (9.1)(a), I_q is prenormal for any such q. This contradicts that (c) holds for Q. We have now proved the claim. We now see that the hypotheses of (9.2) are satisfied, and so the present corollary follows from that theorem.

(9.7) Corollary. Suppose that I is regular, and that U_0 is the set of minimal members of U. Also suppose that $E(I) = A^*(I)$. Then there is a projective extension K of I with $A^*(K) = U \cup E(I)$ if and only if I_Q is not prenormal for every $Q \in U_0$. In particular, the conclusion holds if $A^*(I)$ consists of exactly the primes minimal over I.

Proof. One direction of the first part is by (9.3)(a). The other follows immediately from (9.6), since $U \cap E(I) = \varnothing$ shows that (9.6)(a) holds for all $Q \in U_0$. The last part follows from the first part and (1.1)(e) and (d).

(9.8) Corollary. Let I be regular. The following are equivalent.

i) For every finite set U consisting of primes containing I, there is a projective extension K of I with $A^*(K) = U \cup E(I)$.

ii) For every prime P containing I, there is a projective extension K of I with $P \in A^*(K)$.

iii) For every prime P containing I, either $P \in \overline{A}^*(I)$ or I_P is not prenormal.

Proof. (i) $\Rightarrow$ (ii) is trivial, and (ii) $\Rightarrow$ (iii) is by (9.1)(d). Now suppose

that (iii) holds. We first claim that if $I \subseteq q \in$ Spec R, with I_q prenormal,

then only finitely many primes contain q. Suppose to the contrary, that

infinitely many primes contain q. Then W = $\{P \in$ Spec R $\mid q \subset P$ and

height $P/q = 1\}$ is an infinite set. Let H be as in (9.1)(c). Let

$W' = \{P \in W \mid H \nsubseteq P\}$. Then I_P is prenormal for all $P \in W'$. By (iii),

$W' \subseteq \overline{A}^*(I)$, so that W' is finite. Thus W - W' is infinite, and so

$\cap\{P \in W - W'\} = q$. It follows that $H \subseteq q$, which says I_q is not

prenormal. This contradiction proves the claim. Now let U be any

finite set of primes containing I, harmlessly assuming that

$U \cap E(I) = \emptyset$. Let U_0 be the set of minimal members of U. We will show

that U_0 satisfies the hypotheses of (9.6). If $Q \in U_0$, since U is disjoint

from E(I), and hence from $\overline{A}^*(I)$, (iii) shows that I_Q is not prenormal, as

required by (9.6). Now suppose that $Q \in U_0$ with Q not isolated in U.

Let q be a prime minimal over I with $q \subseteq Q$. Since $q \in E(I)$, $q \subset Q$.
We claim I_q is not prenormal. If it were prenormal, our earlier claim

would show that only finitely many primes contain q. As $q \subset Q$, we

would have to have that Q is a maximal ideal. This contradicts that

$Q \in U_0$ is not isolated in U. This shows that (9.6)(c) holds. We have

now shown that U_0 satisfies the hypotheses of (9.6), and so that result

shows that (i) holds. Thus, (iii) $\Rightarrow$ (i), completing the proof.

So far, we have been considering a projective extension
K of I. If we weaken this to just requiring that K be projectively
equivalent to I, we can say more.

(9.9) Lemma. Let $n \geq 2$, and let $a_1, ..., a_n$ be elements of R. Let $m \geq 2$,
and let $I = (a_1{}^m, ..., a_n{}^m)R$ be a regular ideal. If q is a prime
containing I, and if $a_1, ..., a_n$ are analytically independent in R_q, then
I_q is not prenormal.

Proof. Let $k \geq 1$. We will in fact show that $I_q{}^k$ is not integrally closed.
For this, we can assume that R is local at q. Let $H = (a_1, ..., a_n)$.
Consider the obvious set of generators for H^m. Any element of that set,
when raised to the m, gives a result in I^m. Thus we see that $I \subseteq H^m \subseteq \overline{I}$,

and so $I^k \subseteq H^{mk} \subseteq \overline{I^k}$. If I^k is integrally closed, then $I^k = H^{mk}$.
Since $a_1, ..., a_n$ are analytically independent, H^{mk} is minimally
generated by monomials of degree mk in $a_1, ..., a_n$. Also,
$a_1{}^m, ..., a_n{}^m$ are analytically independent, so I^k is minimally
generated by monomials of degree k in $a_1{}^m, ..., a_n{}^m$. Having
different numbers of minimal generators, (since $n \geq 2$ and $m \geq 2$),
$I^k \neq H^{mk}$, a contradiction.

Recall that I is in the principal class if I can be generated by
n elements with $n = $ height I.

(9.10) Proposition. Let I be a regular ideal in the principal class with
height greater than 1. Then for any finite set U of primes, each of which
contains I, there is an ideal K projectively equivalent to I with
$A^*(K) = U \cup E(I)$.

Proof. Write $I = (a_1, .., a_n)R$ with height $I = n > 1$. For any prime q
containing I, I_q is still in the principal class, and so $a_1, ..., a_n$ are
analytically independent in R_q. By (9.9), $(a_1{}^2, ..., a_n{}^2)R_q$ is not
prenormal. Now (9.8)(iii) $\Rightarrow$ (i) shows there is a projective extension
K of $(a_1{}^2, ..., a_n{}^2)$ with $A^*(K) = U \cup E(I)$. Since $(a_1{}^2, ..., a_n{}^2)$ is
projectively equivalent to I, so is K.

(9.11) Corollary. Let (R, M) be local, with infinite residue field. Let I be
regular, and suppose that the height of I and the analytic spread of I are
equal, and are larger than 1. Then for any finite set U of primes, each
of which contains I, there is an ideal K projectively equivalent to I with
$A^*(K) = U \cup E(I)$.

Proof. Let J be a minimal reduction of I. Then standard facts show that
J is in the principal class, and has the same height as I. The corollary
now follows from (9.10) and the fact that an ideal projectively
equivalent to J is also projectively equivalent to I.

We now turn to the proof of (9.2). We need three lemmas.

(9.12) Lemma. a) Let I be regular. There is an integer $k \geq 1$ with the following property. If for some $n \geq k$, J is an ideal with $J \not\subseteq I^n$, then $(I^n + J)^m \neq (I^n)^m$ for all $m \geq 1$.

b) Let (R, M) be local and let J be an ideal with $I \subseteq J$. If $I^m \neq J^m$ for all large integers m, then $M \in A^*(I + MJ)$.

Proof. a) As is well known, (see [M2, Lemma 8.1]), there is a $k \geq 1$ such that for all $n \geq k$, $(I^{n+h} : I^h) = I^n$ for all $h \geq 0$. Suppose that $n \geq k$ and that $(I^n + J)^m = (I^n)^m$ for some $m \geq 1$. Then $(I^n + J)(I^n)^{m-1} \subseteq (I^n + J)^m = (I^n)^m$. Thus $J \subseteq I^n + J \subseteq (I^{nm} : I^{nm-n}) = I^n$.

b) For m large, by Nakayama's lemma, $J^m \not\subseteq I^m + MJ^m$. Since clearly $(I + MJ)^m \subseteq I^m + MJ^m$, we have $J^m \not\subseteq (I + MJ)^m$. However, $M^m J^m \subseteq (I + MJ)^m$, showing that M^m consists of zero divisors modulo $(I + MJ)^m$. As M is maximal, $M \in \text{Ass } R/(I + MJ)^m$ for all large m, showing $M \in A^*(I + MJ)$.

(9.13) Lemma. i) Let J be an ideal with $J \not\subseteq \cup \{P \in E(I)\}$. Then for all large r, $(I^r : <J>)^n = I^{rn} : <J>$ for all $n \geq 1$. Furthermore, for any prime Q containing J, $Q \notin \text{Ass } R/(I^r : <J>)^n$ for all $n \geq 1$. In particular, $Q \notin A^*(I^r : <J>)$. Also, $I^r : <J>$ is a projective extension of I.

ii) Let J be any ideal. Also, let H be an ideal with $I \subseteq H \subseteq \bar{I}$. Then $(I^r : <J>) \cap H^r$ is a projective extension of I. Also, for large r and all $n \geq 1$, $((I^r : <J>) \cap H^r)^n = (I^{rn} : <J>) \cap H^{rn}$. Also, if $E(I) = A^*(H)$, and if Q is a prime containing I and J with $J_Q \not\subseteq \cup \{P_Q \in E(I_Q)\}$, then for r sufficiently large, $Q \notin \text{Ass } R/((I^r : <J>) \cap H^r)^n$ for all $n \geq 1$. In particular, $Q \notin A^*((I^r : <J>) \cap H^r)$.

Proof. i) Let $R(I, J)$ be as in chapter 5. By (5.3)(a), $R(I, J)$ is a finite $R(I)$-module. By (6.5), for all large r and all $n \geq 1$,

$(I^r : <J>)^n = I^{rn} : <J>$, as desired. Suppose the prime Q contains J. By (1.7)(a), Q is not a prime divisor of $I^{rn} : <J> = (I^r : <J>)^n$ for any $n \geq 1$, and so $Q \notin A^*(I^r : <J>)$, as desired. Also, since R(I, J) is a finite R(I)-module, $R(I, J) \subseteq R(I)' \cap R[u, t] = R[u, \bar{I}t, \overline{I^2} t^2, \overline{I^3} t^3, \cdots]$.

Thus $I^r \subseteq (I^r : <J>) \subseteq \overline{I^r}$, and we are done.

ii) Since $I^r \subseteq (I^r : <J>) \cap H^r \subseteq \overline{I^r}$, the first statement of (ii) is clear. Let $R(I) = R[u, It]$ and $R(H) = R[u, Ht]$ be the Rees rings of R with respect to I and H. As $H \subseteq \bar{I}$, R(H) is contained in the integral closure of R(I), and so R(H) is a finite R(I)-module. Let $B = R[u, I_1 t, I_2 t^2, ...]$, where

$I_r = (I^r : <J>) \cap H^r$. Since $R(I) \subseteq B \subseteq R(H)$, B is a finite R(I)-module. By (6.5), for large r and all $n \geq 1$, $I_r^n = I_{rn}$. This proves the second statement in (ii). Now suppose $E(I) = A^*(H)$, and that Q is a prime containing I and J with $J_Q \not\subseteq \cup\{P_Q \in E(I_Q)\}$. By part (i), for large enough r, $Q_Q \notin$ Ass $R_Q/(I_Q^r : <J_Q>)^n$ for all $n \geq 1$. Thus, it will suffice to show that for sufficiently large r, $I_Q^r : <J_Q> = (I_Q^r : <J_Q>) \cap H_Q^r$. Suppose that r is large enough that $A^*(H) =$ Ass R/H^r. Since $E(I) = A^*(H) =$ Ass R/H^r, we see that $E(I_Q) =$ Ass R_Q/H_Q^r. Thus J_Q does not consist entirely of zero divisors modulo H_Q^r. Since $I_Q^r \subseteq H_Q^r$, this shows that $(I_Q^r : <J_Q>) \cap H_Q^r = I_Q^r : <J_Q>$, as desired.

(9.14) Lemma. Let J be an ideal with $I^n \subseteq J \subseteq \overline{I^n}$, and let a and b be positive integers. If $I^a = J^b$, then $a = nb$.

Proof. $I^a = J^b$ implies that $\overline{I^a} = \overline{J^b}$. However, $I^n \subseteq J \subseteq \overline{I^n}$ implies that $\overline{J^b} = \overline{I^{nb}}$. Thus $\overline{I^a} = \overline{I^{nb}}$. It follows easily from [M2, Lemma 11.27] that $a = nb$.

Definition. Let W be a subset of Spec R. For $P \in$ W, we define

W-rank $P = n$ to mean there is a chain $P_0 \subset P_1 \subset ... \subset P_n = P$ with each P_i in W, but there is no such chain of longer length.

We turn to the main assault.

Proof of (9.2). Let $U_0 = \{Q_1, ..., Q_s\}$, and let $L = I^n + (Q_1 \cdots Q_s) \overline{I^n}$. We will show that n can be chosen large enough that L satisfies the following two conditions.

a) $U_0 \subseteq A^*(L)$.

b) If $Q \in U_0$ and if Q is not isolated in U, then no power of L_Q equals a power of I_Q.

For $Q \in U_0$, since $(Q_1 \cdots Q_s)_Q = Q_Q$, $L_Q = I_Q^n + Q_Q \overline{I_Q^n}$. Since I_Q is not prenormal, (9.1)(a) shows that for all large n, $I_Q^n \neq \overline{I_Q^n}$, so that $\overline{I_Q^n} \not\subseteq I_Q^n$. Thus (9.12)(a) shows that if n is sufficiently large, then for all $m \geq 1$, $(I_Q^n + \overline{I_Q^n})^m \neq (I_Q^n)^m$. That is, $(\overline{I_Q^n})^m \neq (I_Q^n)^m$. By (9.12)(b), we see that $Q_Q \in A^*(L_Q)$. Letting n be large enough that this holds for all $Q \in U_0$, we see that (a) holds. Now suppose that $Q \in U_0$ is not isolated in U. Since for large n we have $Q_Q \overline{I_Q^n} \not\subseteq I_Q^n$, (9.12)(a)) shows that for large n, $(I_Q^n + Q_Q \overline{I_Q^n})^m \neq (I_Q^n)^m$ for all $m \geq 1$. That is, $L_Q^m \neq (I_Q^n)^m$ for all $m \geq 1$. Since $I_Q^n \subseteq L_Q \subseteq \overline{I_Q^n}$, (9.14) shows that no power of I_Q can equal a power of L_Q. This completes the argument that for large enough n, both (a) and (b) hold for L. Note also that L is a projective extension of I.

Let us now give a generic argument which we will use repeatedly, in order to simplify notation. Essentially, we claim that all of the ideals we will be constructing in this proof, can be replaced by any power of themselves. For instance, we have just constructed L, with $I^n \subseteq L \subseteq \overline{I^n}$. Since it does no harm to the theorem to replace I by I^n, we will assume that n = 1, so that $I \subseteq L \subseteq \overline{I}$. Furthermore, we will soon consider a projective extension H of L. Suppose $L^m \subseteq H \subseteq \overline{L^m}$. The reader will have no trouble seeing that the proof is not affected by replacing L by L^m, and so we will do that (also replacing I by I^m). That is, we will let m = 1, and so will simply assume that $I \subseteq L \subseteq H \subseteq \overline{L} = \overline{I}$. In this way, leaving the easy details to the reader, we will always assume that the ideals under consideration fall between I and $\bar{I}$. For the sake of reference, we will call this the power principle.

Let $W = U \cup \{P \in E(I) \mid P$ contains a prime in $U\}$. Let k be the maximum W-rank of a prime in W, and for $0 \le i \le k$, let $W_i = \{P \in W \mid W\text{-rank } P = i\}$. Notice that $W_0 = U_0$, and $W = W_0 \cup \cdots \cup W_k$. We will inductively build projective extensions $L_0, ..., L_k$ of I such that $W_0 \cup ... \cup W_i \subseteq A^*(L_i)$, and such that if $q \in A^*(L_i) - (W_0 \cup ... \cup W_i \cup E(I))$, then q properly contains a prime in W_i. Furthermore, if $Q \in U_0$ is not isolated in U, then no power of $(L_i)_Q$ will equal a power of I_Q.

We now construct L_0. By (8.5) and (8.1), there is a projective extension H of L with $A^*(H) = E(L) = E(I)$. By the power principle discussed previously, we may assume that $I \subseteq L \subseteq H \subseteq \bar{L} = \bar{I}$. Let $L_0 = L + (Q_1 \cdots Q_s)H$ (recalling that $W_0 = U_0 = \{Q_1, ..., Q_s\}$). Since $I \subseteq L \subseteq L_0 \subseteq H \subseteq \bar{I}$, L_0 is a projective extension of I. Suppose that $Q \in U_0$ with Q not isolated in U. We first claim that no power of $(L_0)_Q$ equals a power of I_Q. Since $I_Q \subseteq (L_0)_Q \subseteq \overline{I_Q}$, (9.14) shows it will suffice to prove that $((L_0)_Q)^a \ne I_Q{}^a$, for all $a \ge 1$. If, to the contrary, equality held for some a, then we would have $I_Q{}^a \subseteq L_Q{}^a \subseteq ((L_0)_Q)^a = I_Q{}^a$. Thus $L_Q{}^a = I_Q{}^a$. This contradicts that (b) holds for L, and so proves our first claim. We next claim that $W_0 \subseteq A^*(L_0)$. Since any $Q \in W_0$ is isolated in that set, $(L_0)_Q = L_Q + Q_Q H_Q$. Since (a) holds for L, $Q \in U_0 \subseteq A^*(L)$. This implies that Q_Q is a prime divisor of all big powers of L_Q. On the other hand, $Q \in U$ gives $Q \notin E(I) = A^*(H)$, so that Q_Q is not a prime divisor of any big power of H_Q. Thus for all big m, $L_Q{}^m \ne H_Q{}^m$, so that (9.12)(b) (applied in (R_Q, Q_Q) to $L_Q \subseteq H_Q$) shows $Q_Q \in A^*((L_0)_Q)$. Thus $Q \in A^*(L_0)$, proving our second claim. Our third claim is that if $q \in A^*(L_0) - (W_0 \cup E(I))$, then q properly contains a prime in W_0. Suppose $q \in A^*(L_0)$ and q does not properly contain a prime in W_0. We need $q \in W_0 \cup E(I)$. Suppose $q \notin W_0$. Then q does not contain (properly or improperly) any prime in W_0, and so $(L_0)_q = H_q$. As $q \in A^*(L_0)$, we see that $q \in A^*(H) = E(I)$, as desired. This proves our third claim, and completes the construction of L_0.

Suppose now that $0 \leq i < k$, and that L_i has been constructed.

By the power principal, we may assume that $I \subseteq L_i \subseteq \bar{I}$. We will construct L_{i+1}. First, an auxiliary construction. Let $\{q_1, ..., q_t\} = \{q \in A^*(L_i) - (W_0 \cup ... \cup W_i \cup E(I)) \mid q$ does not contain any prime in $W_{i+1}\}$. Notice that by induction, each q_j properly contains a prime in W_i. Let $J = q_1 \cdots q_t$. By (8.5) applied to L_i there is a projective extension H_i of L_i with $E(I) = E(L_i) = A^*(H_i)$. By the power principle, we may assume that $I \subseteq L_i \subseteq H_i \subseteq \overline{L_i} = \bar{I}$. Let r be sufficiently large, (as in (9.13)(ii) applied to J and $L_i \subseteq H_i \subseteq \overline{L_i}$), and define

$K_i = (L_i{}^r : <J>) \cap H_i{}^r$. By the power principle, we may replace I, L_i, and H_i by their r-th powers, and so get $K_i = (L_i : <J>) \cap H_i$, and $L_i \subseteq K_i \subseteq \bar{I}$. If $P \in W_0 \cup ... \cup W_i$, then P does not contain any q_j (since each q_j properly contains a prime in W_i). Thus, $(K_i)_P = (L_i)_P$. Since by induction $P \in A^*(L_i)$, this shows that $W_0 \cup ... \cup W_i \subseteq A^*(K_i)$. Also, if $Q \in U_0 = W_0$ is not isolated in U, then no power of $(K_i)_Q = (L_i)_Q$ equals a power of I_Q.

Let $W_{i+1} = \{P_1, ..., P_h\}$, and let $L_{i+1} = I + (P_1 \cdots P_h)K_i$. Obviously $I \subseteq L_{i+1} \subseteq \bar{I}$. We claim that $W_0 \cup ... \cup W_{i+1} \subseteq A^*(L_{i+1})$. Let $P \in W_0 \cup ... \cup W_i$. Considering W-ranks, we see that P does not contain any of $P_1, ..., P_h$. Thus $(L_{i+1})_P = (K_i)_P$, and as the previous paragraph shows that $P \in A^*(K_i)$, we get $P \in A^*(L_{i+1})$ as desired. Now let $P \in W_{i+1}$, so that $(L_{i+1})_P = I_P + P_P(K_i)_P$. If $P \in E(I)$, then $P \in A^*(L_{i+1})$ by (8.1) and (1.1)(d). On the other hand, if $P \in W_{i+1} - E(I)$, then P is in U, but is not minimal in U (since $i + 1 > 0$). Thus there is a $Q \in U_0$ with Q properly contained in P, so that Q is not isolated in U. By the previous paragraph, no power of $(K_i)_Q$ equals a power of I_Q. Thus, no power of $(K_i)_P$ equals a power of I_P. By (9.12)(b) (applied in (R_P, P_P) to $I_P \subseteq (K_i)_P$), we see that $P_P \in A^*((L_{i+1})_P)$, so that $P \in A^*(L_{i+1})$, as desired. Next, we claim that if $Q \in U_0$ is not isolated in U, then no power of $(L_{i+1})_Q$ equals a power of I_Q. This is clear, since $(L_{i+1})_Q = (K_i)_Q$.

To complete the construction of L_{i+1}, we must show that if $q \in A^*(L_{i+1}) - (W_0 \cup ... \cup W_{i+1} \cup E(I))$, then q properly contains a prime in W_{i+1}. Suppose not. We will get a contradiction. Since $q \notin W_{i+1}$, q does not contain (properly or improperly) a prime in W_{i+1}, and so $(L_{i+1})_q = (K_i)_q$, showing that $q \in A^*(K_i)$. We claim that q must contain the J used in defining K_i. Since $q \notin W_0 \cup ... \cup W_i \cup E(I)$, and q does not contain any prime in W_{i+1}, the definition of $\{q_1, ..., q_t\}$ shows that either q is in that set, or $q \notin A^*(L_i)$. Thus either our claim is true or $q \notin A^*(L_i)$. Suppose that $q \notin A^*(L_i)$. Since $q \in A^*(K_i)$, we must have that $(K_i)_q \neq (L_i)_q$. This in turn implies that $J \subseteq q$. Therefore, in either case our claim is true. Also note that q contains L_i, since it contains the projectively equivalent ideal L_{i+1}. We now claim that there is a $P \in E(I)$ with $J \subseteq P \subseteq q$. Since $q \in A^*(K_i) = A^*((L_i : <J>) \cap H_i)$, (9.13)(ii) and our choice of r (now set equal to 1) shows that we must have $J_q \subseteq \cup\{P_q \in E(I_q)\}$. Our second claim follows trivially from this. Since $J \subseteq P$, for some $q_j \in \{q_1, ..., q_t\}$ we have $q_j \subseteq P \subseteq q$. Since q_j properly contains a prime in W_i, we see that $P \in W$ and W-rank $P \geq i + 1$. Therefore q contains a prime in W_{i+1}, and that inclusion must be proper, since $q \notin W_{i+1}$. This contradicts our assumption that q does not properly contain a prime in W_{i+1}, and is the contradiction we sought near the start of this paragraph. This completes the construction of L_{i+1}, and hence of our sequence $L_0, ..., L_k$.

We now define the K sought in the statement of the result. Note that $W_0 \cup ... \cup W_k \cup E(I) = U \cup E(I)$. Let D be the product of the primes in $A^*(L_k) - (U \cup E(I))$. Note that $D \not\subseteq P$ for any $P \in U \cup E(I)$, since if $D \subseteq P$, then P would properly contain a prime in W_k, and so we would have $P \in W$ and W-rank $P > k$ which is impossible. In particular, we have $D \not\subseteq \cup\{P \in E(I) = E(L_k)\}$. Let $K = L_k{}^r : <D>$ with r large enough that (9.13)(i) is satisfied. (Here, we will not bother to invoke the power principle.) Thus K is a projective extension of L_k, and hence of I. Let $q \in A^*(K)$. By (9.13)(i), we must have $D \not\subseteq q$, so that $K_q = (L_k)_q{}^r$, showing $q \in A^*(L_k)$. If $q \notin U \cup E(I)$, then q is one of the factors of D, which contradicts that $D \not\subseteq q$. Therefore $q \in U \cup E(I)$. This shows $A^*(K) \subseteq U \cup E(I)$. For the reverse, since K is a projective extension of I, $E(I) \subseteq A^*(K)$. Also, if $P \in U$, we already have $D \not\subseteq P$. Thus $K_P = (L_k)_P^r$. As $P \in W_0 \cup ... \cup W_k$, the construction of L_k shows $P \in A^*(L_k)$, and so $P \in A^*(K)$, as desired. Thus $A^*(K) = U \cup E(I)$.

(9.15) Question. If there is a projective extension L of I with $U \subseteq A^*(L)$, does it follow that there is a projective extension K of I with $A^*(K) = U \cup E(I)$?

10 PRIME DIVISORS OF PRINCIPAL IDEALS

Let $R \subseteq T \subseteq R'$, with the ring T a finite R-module. Let b be a regular element of R, and let P be a prime ideal of R containing b. It may happen that P is a prime divisor of bR, but that no prime divisor of bT lies over P. Conversly, it may happen that P is not a prime divisor of bR, but there is a prime divisor of bT which does lie over P. In essence, the purpose of this chapter is to explore examples of this sort of behavior, except that instead of dealing with a single prime P, we will deal with a finite set of primes.

Let W be a finite set of primes of R, each of which contains the regular element b. We ask what conditions on W will assure that there is a ring T between R and its integral closure, with T a finite R-module, such that $W = \{Q \cap R \mid Q \in \text{Ass } T/bT\}$. By (8.6)(a), we have that for any such T, $E(bR) \subseteq \{Q \cap R \mid Q \in \text{Ass } T/bT\}$, and therefore one necessary condition is that $E(bR) \subseteq W$. We will write $W = U \cup E(bR)$, and assume that $U \cap E(bR) = \varnothing$. We will show that with mild assumptions on U, we can find a T as above with $U \cup E(bR) = \{Q \cap R \mid Q \in \text{Ass} T/bT\}$. This material is taken from [MR3].

Notation. R will be a Noetherian ring with integral closure R'. We will use b to represent a regular element of R, and U will represent a finite set of primes of R each of which contains b. Also, we will assume that $U \cap E(bR) = \varnothing$. $T(R) = \{T \mid T$ is a ring between R and R', with T a finite R-module$\}$.

We relate this topic to the ideas of the previous chapter. Recall that if K is an ideal between bR and $\overline{bR}$, then there is an integer m with $b^n K^m = K^{n+m}$ for all $n \geq 1$.

(10.1) Proposition. Let b be a regular element of R, and let K be an ideal with $bR \subseteq K \subseteq \overline{bR}$. Let m be as described above, and let $T = K^m b^{-m}$. Then $T \in \mathbf{T}(R)$, $T \subseteq R_b$, and the following hold.

i) $bT = KT$, and for all $n \geq m$, $b^n T = K^n T = K^n$.

ii) $A^*(K) = \{Q \cap R \mid Q \in \text{Ass } T/bT\}$.

Proof. Since $b^m R \subseteq K^m \subseteq \overline{b^m R} \subseteq b^m R'$, clearly $T = K^m b^{-m}$ is a finite R-module contained between R and R'. However, $T^2 = K^{2m} b^{-2m} = b^m K^m b^{-2m} = T$, and so T is a ring in $\mathbf{T}(R)$. Clearly $T \subseteq R_b$. Also notice that $K^{m+1} b^{-(m+1)} = bK^m b^{-(m+1)} = K^m b^{-m} = T$, so $T = K^n b^{-n}$ for any $n \geq m$. For such n, clearly $b^n T = K^n$, and as this is an ideal in T, it also equals $K^n T$. Now $(b^{n-1} T)(KT) \subseteq K^n T = b^n T$. Cancelling $b^{n-1} T$ gives $KT \subseteq bT$, and so the two are equal. This proves (i). For (ii), suppose $P \in A^*(K)$. Then for large n, $P \in \text{Ass } R/K^n$. However, if $n \geq m$, then $b^n T \cap R = K^n \cap R = K^n$, and so P lifts to a prime divisor of $b^n T$. As b is regular in T, $b^n T$ and bT have the same prime divisors. This proves half of (ii). Conversly, suppose P lifts to the prime $Q \in \text{Ass } T/bT$. Now $Q = (bT :_T t)$ for some $t \in T$. Intersecting with R, $P = (bT :_R t)$. For all $n \geq m$, we may write $t = x_n b^{-n}$ with $x_n \in K^n$. Thus $P = (b^{n+1} T :_R x_n) = (K^{n+1} :_R x_n)$. As $x_n \in R$, $P \in \text{Ass } R/K^{n+1}$, for all $n \geq m$. Thus $P \in A^*(K)$.

(10.2) Lemma. Let b be a regular element. If for some $n \geq 1$, $b^n R = \overline{b^n R}$ then for any $n \geq m \geq 1$, $b^m R = \overline{b^m R}$. In particular, bR is prenormal if and only if it is normal.

Proof. If $b^n R = \overline{b^n R}$, then $b^n R \subseteq (b^{n-m} R)(\overline{b^m R}) \subseteq \overline{b^n R} = b^n R$, showing that $b^n R = (b^{n-m} R)(\overline{b^m R})$. Cancelling $b^{n-m} R$ gives $b^m R = \overline{b^m R}$. This proves the first statement. The second statement follows easily.

(10.3) Theorem. Let U_0 be the set of minimal members of U, and assume that bRp is not normal for all $P \in U_0$. Also assume that if $P \in U_0$ is not isolated in U, then $Pp(\overline{b^m Rp}) \not\subseteq b^m Rp$ for some $m \geq 1$. Then there is a $T \in \mathbf{T}(R)$ with $T \subseteq R_b$, such that

$U \cup E(bR) = \{Q \cap R \mid Q \in \text{Ass } T/bT\}$.

Proof. By (10.2), bRp is not prenormal for all $P \in U_0$. Also, if $P \in U_0$ is not isolated in U, then there is an $m \geq 1$ with $Pp \; \overline{b^m Rp} \not\subseteq b^m Rp$. It follows easily that for all $n \geq m$, $Pp \; b^n Rp \not\subseteq b^n Rp$, (the argument being similar to the proof of (10.2)). By (9.2), we see that there is a projective extension K of I with $A^*(K) = U \cup E(I)$. Suppose that $b^n R \subseteq K \subseteq \overline{b^n R}$. Applying (10.1) to b^n, we find a $T \in \mathbf{T}(R)$ with $T \subseteq R_{b^n} = R_b$, such that $U \cup E(I) = A^*(K) = \{Q \cap R \mid Q \in \text{Ass } T/b^n T\} = \{Q \cap R \mid Q \in \text{Ass } T/bT\}$, (since Ass $T/b^n T = $ Ass T/bT).

(10.4) Corollary. Let U_0 be the set of minimal members of U, and assume that bRp is not normal for all $P \in U_0$. Also assume that if $P \in U_0$ is not isolated in U, then any one of the following three conditions holds.

a) $P \notin$ Ass R/bR.

b) There is a $q \in$ Ass R/bR - $\overline{A}^*(bR)$ with $b \in q \subset P$.

c) There is a prime q with $b \in q \subset P$ and bR_q not normal.

Then there is a $T \in \mathbf{T}(R)$ with $T \subseteq R_b$, such that $U \cup E(bR) = \{Q \cap R \mid Q \in$ Ass $T/bT\}$.

Proof. The present hypotheses imply that the hypotheses of (10.3) hold, the arguments being similar to those used in the proof of (9.6).

(10.5) Example. Let K be a field and let X and Y be indeterminates. Let R', the integral closure of R, be K[X, Y], while R itself will consist of those polynomials f(X, Y) in R' such that in each monomial of f, neither the exponent of X nor the exponent of Y equals 1. Let $b = X^2Y^2 \in R$. Let $p = XR' \cap R$ and $q = YR' \cap R$. Clearly p and q are the only primes minimal over bR. Since $X^3Y^2 \in \overline{bR_p} - bR_p$, (10.2) shows that bR_p is not prenormal. Similarly, bR_q is not prenormal. Let U be any finite set of primes of R, each of which contains bR, such that $U \cap E(bR) = \emptyset$.

Since $\{p, q\} \subseteq E(I)$, any prime P in U properly contains one of p or q. Since bR_p and bR_q are not prenormal, the same is true of bRp. Also, (10.4)(c) holds. Thus (10.4) implies there is a $T \in \mathbf{T}(R)$ with $T \subseteq R_b$, such that $U \cup E(bR) = \{Q \cap R \mid Q \in \text{Ass } T/bT\}$. (Note: using (1.13)(a) to lift to R', it is not hard to see that $E(bR) = \{p, q\}$.)

(10.6) Theorem. Suppose that every prime in U is isolated in U. The following are equivalent.

a) There is a $T \in \mathbf{T}(R)$, with $T \subseteq R_b$, and with $U \cup E(bR) = \{Q \cap R \mid Q \in \text{Ass } T/bT\}$.

b) There is a $T \in \mathbf{T}(R)$ with $T \subseteq R_b$, such that $U \subseteq \{Q \cap R \mid Q \in \text{Ass } T/bT\}$.

c) For each $P \in U$, there is a $T \in \mathbf{T}(R)$, with $T \subseteq R_b$, such that P lifts to a prime in Ass T/bT.

d) bRp is not normal for all $P \in U$.

Proof. a) $\Rightarrow$ b) $\Rightarrow$ c) are trivial. Suppose that (c) holds. Let $P \in U$. Then there is a $T \in \mathbf{T}(R)$ with $T \subseteq R_b$ such that P lifts to a prime $Q \in \text{Ass } T/bT$. To show (d), we need that bRp is not normal. Let us assume to the contrary that bRp is normal. We may localize at R - P, and assume that P is maximal. Since $T \subseteq R_b$, for some $n \geq 1$ we have $b^n T \subseteq R$. Now $Q \in$ Ass $T/bT =$ Ass $T/b^n T$. Write $Q = (b^n T :_T t)$ with $t \in T$. We may write $t = c/b^n$, with $c \in R$, and so $Q = (b^{2n} T :_T c)$. Now $b^{2n} R \subseteq b^{2n} T \subseteq b^{2n} R' \cap R = \overline{b^{2n} R} = b^{2n} R$ (since we are assuming bR to be normal). Thus $b^{2n} T = \overline{b^{2n} R}$, and so $Q = (\overline{b^{2n} R} :_T c)$. Intersecting with R, we get $P = (\overline{b^{2n} R} :_R c)$. By (1.2) and (1.1)(d), $P \in A^*(bR) \subseteq E(bR)$. This contradicts that $U \cap E(bR) = \emptyset$.

Thus (c) $\Rightarrow$ (d). Finally, (d) $\Rightarrow$ (a) is trivial from (10.3) and the fact that every $P \in U$ is isolated in U.

Remark. The implication (c) $\Rightarrow$ (b) in (10.6) does not hold without the assumption that every prime in U is isolated in U. We will give an example in (10.8).

In the next result, we do not start with a given element b. Recall that $\mathbf{P}(R)$ and $\mathbf{E}(R)$ were defined in chapter 5.

(10.7) Proposition. Let W be a finite set of regular primes of R. Then there is a $T \in \mathbf{T}(R)$ with the following property. If Q is prime in T and $Q \cap R \in W$, then either $Q \in \mathbf{P}(T)$, or T_Q is normal.

Proof. As only finitely many primes of R' lie over any given prime in R, we may easily construct an $R_1 \in \mathbf{T}(R)$ such that if $W_1 = \{Q$ prime in $R_1 \mid Q \cap R \in W\}$ then each $Q \in W_1$ has only one prime of R' lying over it. By replacing R and W by R_1 and W_1, we may assume that if $P \in W$ and $T \in \mathbf{T}(R)$, then P lifts to a unique prime of T. We will use W_T to denote the set $\{Q$ prime in $T \mid Q \cap R \in W\}$. Let $W' = \{P \in W \mid$ there is an $A \in \mathbf{T}(R)$ such that P lifts to the (unique) prime Q of A, and A_Q is normal$\}$. If $A \subseteq B$ are in $\mathbf{T}(R)$ and the prime Q of A lifts to the prime q of B, and A_Q is normal, then B_q is also normal. Using this, (and replacing R by an approriate ring in $\mathbf{T}(R)$) we may asssume that for any $C \in \mathbf{T}(R)$, if $Q \in W'_C$, then C_Q is normal, while if $Q \in (W - W')_C$, then C_Q is not normal. Clearly it does no harm to assume that W' is empty. Thus, we are assuming that for any $C \in \mathbf{T}(R)$, and $Q \in W_C$, C_Q is not normal. Our goal now is to find a $T \in \mathbf{T}(R)$ such that $W_T \subseteq \mathbf{P}(T)$. By (8.6)(a), it is easily seen that any prime in $\mathbf{E}(R)$ lifts to a prime in $\mathbf{P}(T)$ for any $T \in \mathbf{T}(R)$, and so we may assume that $W \cap \mathbf{E}(R) = \varnothing$. Let $P \in W$. We claim that $R'_{(R-P)}$ is not a finite R_P-module. If this were false, we could easily find a $C \in \mathbf{T}(R)$ and a prime Q of C lying over P, such that $C_Q = R'_{(R-P)}$. However, since $Q \in W_C$, this would contradict that C_Q is not normal, and so our claim is true. For $P \in W$, this claim shows that $P_P R'_{(R-P)} \nsubseteq R_P$, and so we can find an $\alpha \in R'$ whose image in $R'_{(R-P)}$ is not in $(R_P : P_P)$. For each $P \in W$, find such an α, and let S be the ring generated over R by those finitely many α. Clearly $S \in \mathbf{T}(R)$, and so there is a regular element b in R with $bS \subseteq R$. We will apply (10.3) to b and $U = W$. Since $W \cap \mathbf{E}(R) = \varnothing$, $W \cap \mathbf{E}(bR) = \varnothing$. If $P \in W$, then let α be the corresponding element chosen above. Since $\alpha \notin R_P$, we have $b \in P$. Now $b\alpha \in bR' \cap R = \overline{bR}$, so that $P_P(b\alpha) \subseteq P_P(\overline{bR_P})$. However, $P_P(b\alpha) \nsubseteq bR_P$, since $P_P\alpha \nsubseteq R_P$. Thus $P_P(\overline{bR_P}) \nsubseteq bR_P$. Therefore, b and W satisfy the hypothesis of (10.3), which shows there is a $T \in \mathbf{T}(R)$ with $W \subseteq \{Q \cap R \mid Q \in \text{Ass } T/bT\}$. Thus, $W_T \subseteq \text{Ass } T/bT$. However, Ass $T/bT \subseteq \mathbf{P}(T)$, and we are done.

We have mentioned that (iii) $\Rightarrow$ (ii) of (9.4), and (c) $\Rightarrow$ (b) of (10.6) do not hold in general. We now give an example of their failure.

(10.8) Example. We will construct a regular principal ideal $I = bR$, and a set $U = \{P_1, P_2\}$ of primes, with $I \subseteq P_1 \subset P_2$, such that for $i = 1, 2$, there is a projective extension K_i of I with $P_i \in A^*(K_i)$, but such that there is no K projectively equivalent to I (and hence no projective extension K of I) with $\{P_1, P_2\} \subseteq A^*(K)$. Thus it will show that in (9.4)(iii) $\Rightarrow$ (ii), we need that the primes in U are all isolated in U. Also, for i = 1, 2, we will show there are rings $T_i \in \mathbf{T}(R)$, with $T_i \subseteq R_b$, such that P_i lifts to a prime $Q_i \in$ Ass T_i/bT_i, but that there is no $T \in \mathbf{T}(R)$ such that both P_1 and P_2 lift to primes in $\mathbf{P}(T)$. This shows that in (10.6)(c) $\Rightarrow$ (b), that same restriction on U is needed.

Let (R, M) be a local domain whose integral closure $R' \neq R$ is a finite R-module, and has R/M infinite. Let $M \notin \mathbf{P}(R)$, let $P \in \mathbf{P}(R)$, and assume that for any ring $T \in \mathbf{T}(R)$ with $T \neq R$, P does not lift to a prime in $\mathbf{P}(T)$. (In the final paragraph, we will show that this is all easily accomplished.) We will take $P_1 = P$ and $P_2 = M$. Also, we will let b be any nonzero element of $P \cap (R : R')$. Note that since $b \in P \in \mathbf{P}(R)$, we must have $P \in$ Ass $R/bR = A^*(bR)$. Thus if $K_1 = bR$, then $P \in A^*(K_1)$, and if $T_1 = R$, then $T_1 \in \mathbf{T}(R)$, $T_1 \subseteq R_b$, and P lifts to a prime in Ass T_1/bT_1. Also, there is a projective extension K_2 of bR with $M \in A^*(K_2)$, and a $T_2 \in \mathbf{T}(R)$, with $T_2 \subseteq R_b$, such that P_2 lifts to a prime in Ass T_2/bT_2. By (10.1), the former statement implies the latter. To prove the former, since $R \neq R'$, $bR \neq bR' = bR' \cap R = \overline{bR}$. Suppose now that for some $m \geq 1$, $(bR)^m = (\overline{bR})^m$. Then $\overline{bR} \subseteq ((bR)^m : (bR)^{m-1}) = bR$. This leads to the contradiction that $bR = \overline{bR}$. Thus, $(bR)^m \neq (\overline{bR})^m$ for all $m \geq 1$. Now (9.12)(b) shows that $M \in A^*(bR + M(\overline{bR}))$. Thus, we let $K_2 = bR + M(\overline{bR})$.

We now claim that there is no $T \in \mathbf{T}(R)$ such that both P and M lift to primes in $\mathbf{P}(T)$. If such a T exists, then by choice of P, we must have $T = R$, which contradicts that M is not in $\mathbf{P}(R)$. We finally claim that there is no ideal K projectively equivalent to bR, with $\{P, M\} \subseteq A^*(K)$. Suppose to the contrary that such a K exists. Then the analytic spread of K is 1, and so we let cR be a minimal reduction of K. Thus $cR \subseteq K \subseteq \overline{cR}$. By (10.1), there is a $T \in \mathbf{T}(R)$ with $\{P, M\} \subseteq A^*(K)$ $= \{Q \cap R \mid Q \in$ Ass $T/cT\}$. Thus, both P and M lift to primes in $\mathbf{P}(T)$. We have just seen that this cannot be.

We now construct R, M, and P, satisfying the assumptions made above. It is easy to find a Noetherian domain A whose integral closure A' is a finite A-module, and a depth ≥ 2 prime Q in $\mathbf{P}(A)$ such that no height 1 prime of A' lies over Q. Using the ascending chain condition for A-submodules of A', let R be a ring between A and A' such that R is maximal with respect to the property that Q lifts to a prime P in $\mathbf{P}(R)$. Thus, if $T \in \mathbf{T}(R)$ with $T \neq R$, then P does not lift to a prime in $\mathbf{P}(T)$, as desired. Since A' is a Krull domain, primes in $\mathbf{P}(A')$ have height 1, so that Q does not lift to a prime in $\mathbf{P}(A')$. Thus $R \neq A' = R'$, as desired. Since depth $Q \geq 2$, depth $P \geq 2$. Infinitely many primes $M \supset P$ have height $M/P = 1$ and depth $M \geq 1$. If such an M is in $\mathbf{P}(R)$, and if $0 \neq x \in P$, then $M \in \text{Ass } R/xR$, which is a finite set. Thus, we can pick our M with $M \notin \mathbf{P}(R)$. As depth $M \geq 1$, R/M is infinite (since finite domains are fields). We localize R at this M.

11 SPORADIC PRIMES

The sequence Ass R/I, Ass R/I^2, Ass R/I^3, $\cdots$ eventually stabilizes to $A^*(I)$, so that Ass $R/I^n = A^*(I)$ for all large n. However, for small n it may happen that there are prime ideals P with $P \in$ Ass $R/I^n - A^*(I)$. We call such primes the sporadic primes of I.

Notation. For all $n \geq 1$, $S_n(I) =$ Ass $R/I^n - A^*(I)$, and $S(I) = \cup S_n(I)$ over all $n \geq 1$. Thus $S(I)$ is the set of sporadic primes of I. Throughout this chapter, I will be regular.

This chapter, taken from [MR4], concentrates on $S_1(I)$. Let W be a finite set of primes, each of which contains I. Our goal will be to find conditions on W which assure the existence of an ideal K, projectively equivalent to I, such that $S_1(K) = W$ while $A^*(K) = A^*(I)$. That is, we want to preserve the persistent primes of I, while gaining some control over the sporadic primes. The next definition will be crucial to us.

Definition. For I a regular ideal, I^* will denote the eventual stable value of $(I^2 : I) \subseteq (I^3 : I^2) \subseteq (I^4 : I^3) \subseteq \cdots$.

(11.1) Proposition. i) If J is an ideal containing I, then $J \subseteq I^*$ if and only if $J^m = I^m$ for some $m \geq 1$ if and only if $J^m = I^m$ for all large m. In particular, if $I \subseteq J \subseteq I^*$, then $A^*(I) = A^*(J)$.

ii) If J is a projective extension of I, with $I^n \subseteq J \subseteq \overline{I^n}$, then $J \subseteq (I^n)^*$ if and only if some power of J equals a power of I.

iii) $I \subseteq I^* \subseteq \bar{I}$.

iv) For all large n, $(I^n)^* = I^n$.

v) For $n \geq 1$, $(I^{n+1} : I) \subseteq (I^{n+2} : I^2) \subseteq (I^{n+3} : I^3) \subseteq \cdots$ eventually stabilizes to $(I^n)^*$.

vi) For $0 \leq m \leq n$, $(I^{n*} : I^{m*}) = (I^{n*} : I^m) = (I^{n-m})^*$.

vii) Ass $R/I^* \subseteq$ Ass $R/I^{2*} \subseteq$ Ass $R/I^{3*} \subseteq \cdots$, and this increasing sequence eventually stabilizes to $A^*(I)$.

viii) Ass $\bar{I}/I^* \subseteq$ Ass $\overline{I^2}/I^{2*} \subseteq$ Ass $\overline{I^3}/I^{3*} \subseteq \ldots$ is increasing, and eventually stabilizes to a subset 0f $A^*(I)$ (denoted $C^*(I)$).

ix) For $n \geq 1$, $S_n(I) \subseteq$ Ass I^{n*}/I^n.

Proof (iii), (iv) and the first part of (i) are proved in [M2, Lemma 8.2]. The last sentence in (i) follows from the rest of (i).

ii) Let $I^n \subseteq J \subseteq \overline{I^n}$. If $I^a = J^b$, then by (9.14), a = nb. Thus, $(I^n)^b = J^b$, so that (i) shows $J \subseteq (I^n)^*$. Conversly, if $J \subseteq (I^n)^*$, then (i) shows that some power of I equals a power of J.

v) The chain given in (v) contains $(I^{(k+1)n} : I^{kn})$ which equals I^{n*} for all large k. The result follows at once.

vi) By (v), let k be large enough that $(I^{n+k} : I^k) = I^{n*}$. Now an easy exercise using (i) shows that $I^{m*}(I^{n-m})^* \subseteq I^{n*}$. Thus $(I^{n-m})^* \subseteq (I^{n*} : I^{m*}) \subseteq (I^{n*} : I^m) = ((I^{n+k} : I^k) : I^m) = (I^{n+k} : I^{m+k}) \subseteq (I^{n-m})^*$, the last inclusion by (v). Equality holds throughout.

vii) Let $P \in$ Ass R/I^{n*}. We want $P \in$ Ass $R/(I^{n+1})^*$. It does no harm to assume that R is local at P. Write $P = (I^{n*} : c)$ with $c \in R - I^{n*}$. Thus $P \subseteq ((I^{n*})I^* : cI^*) \subseteq ((I^{n+1})^* : cI^*)$. Since $c \notin I^{n*} = ((I^{n+1})^* : I^*)$ (by (vi)), $((I^{n+1})^* : cI^*)$ is a proper ideal, and hence contained in P. Thus it equals P, and so $P \in$ Ass $R/(I^{n+1})^*$. This proves the first part of (vii). The second part is trivial, using (iv).

viii) The first part of the proof is identical to that of (vii) with the added observation that if $c \in \overline{I^n}$, then $cI^* \subseteq \overline{I^{n+1}}$. The second part follows from the finiteness of $A^*(I)$ and the fact that

Ass $\overline{I^n}/I^{n^*} \subseteq$ Ass $R/I^{n^*} \subseteq A^*(I)$.

ix) Suppose $P \in S_n(I)$. We want $P \in$ Ass I^{n^*}/I^n. Assume that R is local at P. Write $P = (I^n : c)$. We can finish by showing that $c \in I^{n^*}$. If not, then $(I^{n^*} : c)$ is proper and hence contained in P. Thus

$P = (I^n : c) \subseteq (I^{n^*} : c) \subseteq P$, so that $(I^{n^*} : c) = P$, and $P \in$ Ass R/I^{n^*}.

By (vii), $P \in A^*(I)$, contradicting $P \in S_n(I)$.

(11.2) Remarks: a) Concerning $C^*(I)$ (as in (viii)), by (iv),

$C^*(I) =$ Ass $\overline{I^n}/I^n$ for all large n. For more on $C^*(I)$ and its connection to prenormality, see [M2, Propositions 11.17, 11.18, and 11.19].)

b) Suppose that (R, M) is local and I is M-primary. Then I^* is the unique largest ideal containing I having the same Hilbert polynomial as I. In [Sh], K. Shah extends this concept to find, for an M-primary ideal I in a quasi-unmixed local ring (R, M) with infinite residue field, ideals $I_d = I^*, I_{d-1}, ..., I_0 = \overline{I}$, such that I_k is the unique largest ideal containing I such that I and I_k have their first $k + 1$ Hilbert coefficients in common.

(11.3) Example. Let F be a field and let X and Y be indeterminates. Let $R = F + (X^3, Y)F[X, Y]$, let $P = (X^3, Y)F[X, Y]$, and let $I = (X^3, X^4)R$. Since $X^3 R \subseteq I \subseteq (X^3, X^4, X^5)R = (X^3 R)F[X, Y] \cap R = \overline{X^3 R}$, we have $\overline{I} = (X^3, X^4, X^5)R \neq I$. However, $I^2 = (X^6, X^7, X^8)R = \overline{I}^2$, and so (11.1)(i) and (iii) show that $I \neq I^* = \overline{I}$. For $n \geq 2$, we easily see that $I^n = (X^{3n}, X^{3n+1}, X^{3n+2})R = \overline{I^n} = \overline{X^{3n}R}$. By (11.1)(iii), $I^n = I^{n^*} = \overline{I^n}$

for $n \geq 2$. Now $P = (I : X^5)$, so that $P \in$ Ass R/I. On the other hand, if for $n \geq 2, P \in$ Ass $R/I^n =$ Ass $R/\overline{X^{3n}R} =$ Ass $R/(X^{3n}F[X, Y] \cap R)$, then P would lift to a prime divisor of $X^{3n}F[X, Y]$. This means that P would lift to a height 1 prime of $F[X, Y]$. As it does not, we have $P \notin$ Ass R/I^n for all $n \geq 1$. Therefore, $P \in S_1(I)$. (This comes from [B].)

(11.4) Lemma. Let K be an ideal with $I \subseteq K \subseteq I^*$, and let $P \in S_1(K)$. Then $P \notin A^*(I)$, and $(I : I^*) \subseteq P$.

Proof. Since $P \in S_1(K)$, $P \notin A^*(K)$. Now (11.1)(i) shows that $A^*(K) = A^*(I)$, and so $P \notin A^*(I)$, as desired. Next, suppose that $(I : I^*) \not\subseteq P$. Then it is easily seen that $I_P = (I^*)_P = (I_P)^*$. As $I \subseteq K \subseteq I^*$, we must have $K_P = I_P^*$. Since $P \in S_1(K)$, $P \in$ Ass R/K. Therefore, $P_P \in$ Ass $R_P/K_P =$ Ass $R_P/I_P^* \subseteq A^*(I_P)$ (using (11.1)(vii)). This shows that $P \in A^*(I)$, which contradicts what we have just proved.

We now give the main result of this chapter. Of the three conditions imposed below on W, the need for two of them is explained by (11.4). The third condition, (the most cumbersome of them), is needed to make an induction work.

(11.5) Theorem. Suppose that $I \neq I^*$. Let W be a finite set of primes of R with $W \cap A^*(I) = \varnothing$, and with $(I : I^*) \subseteq P$ for all $P \in$ W. Furthermore, suppose that if $P \in$ W satisfies $(I : I^*)_P = P_P$, then P is isolated in W. Then there is an ideal K with $I \subseteq K \subseteq I^*$, and $S_1(K) = W$.

In order to prove (11.5), we need the following lemma.

(11.6) Lemma. Let $I \subseteq H$ be ideals of R. Let $V = \{P_1, ..., P_n\}$ be a finite set of primes of R such that each $P \in$ V is isolated in V. Suppose that $I_P \neq H_P$ for all $P \in$ V. Let $J = I + P_1...P_nH$. Then

a) $V \subseteq$ Ass R/J.

b) If $Q \in S_1(J)$, then either Q contains some $P \in$ V or $Q \in S_1(H)$.

Proof. Let $P \in$ V. Since P is both maximal and minimal in V, $J_P = I_P + P_PH_P$. Since $I_P \neq H_P$, Nakayama's lemma shows that

$J_P \neq H_P$. Thus $(J_P : H_P)$ is a proper ideal in R_P, and so is contained in P_P. However, it clearly contains P_P, so that $P_P = (J_P : H_P)$.

Thus $P_P \in$ Ass R_P/J_P, and so $P \in$ Ass R/J, proving (a). For (b), suppose that $Q \in S_1(J)$, and $P \not\subseteq Q$ for all $P \in V$. Then $J_Q = H_Q$. Since $Q \in$ Ass $R/J - A^*(J)$, we have that $Q_Q \in$ Ass $R_Q/J_Q - A^*(J_Q) =$ Ass $R_Q/H_Q - A^*(H_Q)$. Thus $Q \in S_1(H)$. (Remark: In (11.6), we can drop the assumption that every prime in V is isolated in V, if we add the assumption that if P is minimal but not isolated in V, then $P_P H_P \not\subseteq I_P$. [MR4, (2.5)].)

Proof of (11.5). Recall that W-rank was defined following (9.14). Let r be the maximum W-rank of a prime in W, and for i = 0, 1, ..., r, let $W_i = \{P \in W \mid$ W-rank $P = i\}$. Also, let $W_{-1} = \varnothing$. We will inductively construct ideals $I_{-1}, I_0, ..., I_r$, between I and I^*, such that for $-1 \leq i \leq r$, $S_1(I_i) = W_{-1} \cup W_0 \cup ... \cup W_i$. The result will follow from taking $K = I_r$. In our construction, we will have that $(I_i)_q \neq I_q$, for all $-1 \leq i \leq r$ and all $q \in W_0$ with q not isolated in W. (This will be needed to keep the induction going.)

We start by letting $I_{-1} = I^*$. Note that (11.1)(i) shows $A^*(I) = A^*(I^*)$. Thus (11.1)(vii) shows that Ass $R/I^* \subseteq A^*(I^*)$, so that $S_1(I_{-1}) = \varnothing = W_{-1}$. Also, by hypothesis, $(I : I^*) \subseteq P$ for all $P \in W$, so in particular $(I_{-1})_q \neq I_q$ for all $q \in W_0$ with q not isolated in W. Thus I_{-1} is as desired.

Suppose for some $0 \leq i \leq r$, we have I_{i-1}. Before constructing I_i, we need an auxilliary ideal. Let $J_i = I + w_i I_{i-1}$ with w_i the product of the primes in W_i. We will show that $W_{-1} \cup ... \cup W_i \subseteq S_1(J_i)$. First note that since by induction, $I \subseteq I_{i-1} \subseteq I^*$, we have $I \subseteq J_i \subseteq I^*$. Thus $A^*(J_i) = A^*(I)$, and so the hypothesis that $W \cap A^*(I) = \varnothing$ shows that $W_{-1} \cup ... \cup W_i$ is disjoint from $A^*(J_i)$. Therefore, we must only show that $W_{-1} \cup ... \cup W_i \subseteq$ Ass R/J_i. We start with any $Q \in W_{-1} \cup ... \cup W_{i-1}$ (there being no such Q if i = 0). Clearly Q cannot contain any prime in W_i, and so $(w_i)_Q = R_Q$. Thus $(J_i)_Q = (I_{i-1})_Q$.

By induction, $W_{-1} \cup \ldots \cup W_{i-1} = S_1(I_{i-1}) \subseteq \text{Ass } R/I_{i-1}$. Therefore

$Q_Q \in \text{Ass } R_Q/(I_{i-1})_Q = \text{Ass } R_Q/(J_i)_Q$, so that $Q \in \text{Ass } R/J_i$, as desired.

It remains to show that $W_i \subseteq \text{Ass } R/J_i$. We claim that for $P \in W_i$,

$(I_{i-1})_P \neq I_P$. If instead $(I_{i-1})_P = I_P$, then let $q \subseteq P$ with $q \in W_0$.

Localizing at qp gives $(I_{i-1})_q = I_q$. If $i = 0$, then this says that $I^*_q = I_q$,

which contradicts the hypothesis that $(I : I^*) \subseteq q$. If $i > 0$, then $q \neq P$, and

so q is not isolated in W. Part of our inductive assumption on I_{i-1} says

that $(I_{i-1})_q \neq I_q$, giving a contradiction. This proves the claim.

By (11.6)(a) we now see that $W_i \subseteq \text{Ass } R/J_i$. This completes the

argument that $W_{-1} \cup \ldots \cup W_i \subseteq \text{Ass } R/J_i$.

Let $U_i = S_1(J_i) - (W_{-1} \cup \ldots \cup W_i)$, and if $U_i \neq \varnothing$, let $P' \in U_i$.

We claim that P' properly contains a prime in W_i. If not, then since

$P' \notin W_i$, P' does not contain (properly or improperly) any prime in W_i.

By (11.6)(b), $P' \in S_1(I_{i-1}) = W_{-1} \cup \ldots \cup W_{i-1}$ (by induction). This

contradicts that $P' \in U_i$.

We now define $I_i = (J_i : \langle u_i \rangle) \cap I^*$, with u_i the product of the

primes in U_i. (If U_i is empty, $u_i = R$.) Clearly $I \subseteq I_i \subseteq I^*$, as desired.

Now let $q \in W_0$ with q not isolated in W. We claim that $(I_i)_q \neq I_q$ (as is

required in the first paragraph). Note that q cannot contain any

$P' \in U_i$, since such a P' has been shown to properly contain a prime in

W_i, and q is minimal in W. Thus $(u_i)_q = R_q$, and $(I_i)_q = (J_i)_q$.

However, $J_i = I + w_i I_{i-1}$. If $i > 0$, then $(w_i)_q = R_q$, and $(J_i)_q = (I_{i-1})_q \neq I_q$

by induction. On the other hand, if $i = 0$, (so that $I_{i-1} = I^*$) then q is one of

the factors of w_i, and so $(J_i)_q = I_q + q_q I^*_q$. Since $q \in W$ but q is not

isolated in W, by assumption we have both $(I : I^*) \subseteq q$, and $(I : I^*)_q \neq q_q$.

Therefore, $(I : I^*)_q \subset q_q$ and so we see that $q_q I^*_q \nsubseteq I_q$.

Thus $(I_i)_q = (J_i)_q = I_q + q_q I^*_q \neq I_q$, as claimed.

To complete the proof, it remains to show that

$S_1(I_i) = W_{-1} \cup \ldots \cup W_i$. Let $P \in W_{-1} \cup \ldots \cup W_i$. Considering

W-ranks, we see that P does not contain any prime P' in U_i, since we

know that such a P' must properly contain a prime in W_i. Therefore,

$(I_i)_P = (J_i)_P$. Since $P \in W_{-1} \cup ... \cup W_i \subseteq S_1(J_i)$, we easily see that

$P \in S_1(I_i)$. Thus, $W_{-1} \cup ... \cup W_i \subseteq S_1(I_i)$. Conversly, let $Q \in S_1(I_i)$.

Then $Q \in$ Ass R/I_i. Now a primary decomposition of I_i can be obtained
by intersecting a primary decomposition of $J_i : <u_i>$ with primary a
decomposition of I^*, and then deleting redundancies. Thus Q must
appear either in Ass $R/(J_i : <u_i>)$ or in Ass R/I^*. Suppose

$Q \in$ Ass R/I^*. Then (11.1)(vii) shows that $Q \in A^*(I) = A^*(I_i)$, the

equality by (11.1)(i). This contradicts that $Q \in S_1(I_i)$. Therefore, we

must have $Q \in$ Ass $R/(J_i : <u_i>)$. By (1.7)(a), we must have $u_i \nsubseteq Q$.

Therefore, $(I_i)_Q = (J_i)_Q$. As $Q \in S_1(I_i)$, we get

$Q \in S_1(J_i) = U_i \cup W_{-1} \cup ... \cup W_i$. However, Q cannot be in U_i, since

we already have $u_i \nsubseteq Q$. Thus $Q \in W_{-1} \cup ... \cup W_i$ as desired.

(11.7) Lemma. Let $J \subseteq L$ be a regular ideals of R with J a reduction of L,
but with $JL \neq L^2$. Let $I = JL$. Then $I \neq I^*$

Proof. Since J reduces L, for some n, $JL^n = L^{n+1}$. By hypothesis, we
must have $n \geq 2$. We get $I^n = J^n L^n = L^{2n} = (L^2)^n$. By (11.1)(i),
$I \subseteq L^2 \subseteq I^*$. Since $I = JL \neq L^2$, $I \neq I^*$.

(11.8) Lemma. Let $H = (a_1, ..., a_n)$ be height $n \geq 2$ regular ideal.
Let $I = (a_1{}^3, ..., a_n{}^3)(a_1{}^3, ..., a_n{}^3, a_1{}^2 a_2)$. Then $I \neq I^*$, and I is
projectively equivalent to H.

Proof. Let $J = (a_1{}^3, ..., a_n{}^3)$, and $L = (J, a_1{}^2 a_2)$. Thus $I = JL$. Since

$J \subseteq L \subseteq H^3 \subseteq \overline{H^3} = \overline{J}$, we see that J and L are projectively equivalent to

H, so I is as well. To show that $I \neq I^*$, we in fact show more, namely that
$I_P \neq I_P{}^*$ for any prime P containing I (equivalently, P containing H).
Note that localizing at such a P does not affect the hypothesis, and so we
may assume that R is local. That height $(a_1, ..., a_n) = n$ now implies
that $a_1, ..., a_n$ are analytically independent. Since we already have

that $J \subseteq L \subseteq \overline{J}$, J is a reduction of L. By (11.7), it will suffice to show that
$JL \neq L^2$. By analytic independence, we easily see that
$(a_1{}^2 a_2)^2 \in L^2 - JL$.

(11.9) Example. Let $R = F[X, Y, Z_1, ..., Z_n]$ with F a field,
$X, Y, Z_1, ..., Z_n$ indeterminates, and $n \geq 1$. Let $Q = (X, Y)R$, and let W
be any finite set of primes of R such that each P in W properly contains
Q. We will show that there is an ideal K projectively equivalent to Q
with Ass $R/K = W \cup \{Q\}$, and with Ass $R/K^2 = $ Ass $R/K^3 = ... = \{Q\}$.
(Thus $S_1(K) = S(K) = W$.) Let $T = F[X, Y]$, let $q = (X, Y)T$, and let
$I_0 = (X^3, Y^3)(X^3, Y^3, X^2Y)T$. By (11.8), $I_0 \neq I_0^*$, and I_0 is projectively
equivalent to q. By (11.1)(iv), we may pick m to be the largest integer
with $I_0^m \neq I_0^{m*}$. Since $(I_0^m : I_0^{m*})$ is proper, and q is the only prime of
T containing I_0^m, q is the unique prime minimal over $(I_0^m : I_0^{m*})$.
Also, we easily see that Ass $T/I_0^n = \{q\}$ for $n \geq 1$.

        Let $I = I_0R$. As R is a faithfully flat extension of T, we see
that $I^{m*} = I_0^{m*}R$, and m is the largest integer such that $I^m \neq I^{m*}$.
Since $(I^m : I^{m*}) = (I_0^m : I_0^{m*})R$, we see that $Q = qR$ is the only prime
minimal over $(I^m : I^{m*})$. Also, Ass $R/I^n = \{Q\}$ for all $n \geq 1$.

        Now let W be any finite set of primes of R such that
P properly contains Q for each $P \in W$. We claim that the hypotheses of
(11.5) hold for W and the ideal I^m. We already have $I^m \neq I^{m*}$. Also,
since $A^*(I^m) = \{Q\}$, $W \cap A^*(I^m) = \emptyset$. Furthermore, we know that
$(I^m : I^{m*}) \subseteq Q \subset P$ for all $P \in W$. Finally, since no $P \in W$ is minimal
over $(I^m : I^{m*})$, clearly $(I^m : I^{m*})_P \neq P_P$. This shows that (11.5) holds
for I^m and W. Thus there is an ideal K with $I^m \subseteq K \subseteq I^{m*}$ and with
$S_1(K) = W$. For $n \geq 2$, $I^{mn} \subseteq K^n \subseteq I^{mn*} = I^{mn}$ (by choice of m), so that
Ass $R/K^n = $ Ass $R/I^{mn} = \{Q\}$. Since clearly $Q \in$ Ass R/K, the definition
of $S_1(K)$ shows that Ass $R/K = W \cup \{Q\}$. Finally, since I_0 is projectively
equivalent to q, I is projectively equivalent to Q, and so K is projectively
equivalent to Q.

12 IRRELEVANT PRIME DIVISORS OF uR(I)

Notation. I will always be a regular ideal in the Noetherian ring R, and P will be a prime ideal of R with $I \subseteq P$.

Definition. The prime ideal Q in R(I), the Rees ring of R with respect to I, is called irrelevant if it contains It. Otherwise, Q is relevant. (Notice that if Q is a prime divisor of uR(I), then Q is homogeneous, and so Q is irrelevant if and only if $Q = (u, P, It)$, where $P = Q \cap R$.)

Our first result shows the relevence of relevant and irrelevant prime divisors of uR(I) to ideas we have previously discussed.

(12.1) Theorem. i) $P \in \text{Ass } R/I^n$ for some $n \geq 1$ if and only if there is a prime divisor Q of uR(I) with $Q \cap R = P$.

ii) $P \in A^*(I)$ if and only if there is a relevant prime divisor Q of uR(I) with $Q \cap R = P$.

iii) $P \in S(I)$ if and only if (u, P, It) is the only prime divisor of uR(I) which lies over P.

Proof. i) Let $P \in \text{Ass } R/I^n$. Since $u^n R(I) \cap R = I^n$, P lifts to a $Q \in \text{Ass } R(I)/u^n R(I)$. Since u is regular in R(I), $Q \in \text{Ass } R(I)/uR(I)$. Conversly, let $Q \in \text{Ass } R(I)/uR(I)$ with $Q \cap R = P$. There is a homogeneous element ct^{m-1}, $c \in I^{m-1}$, with $Q = (uR(I) : ct^{m-1})$. Note that $m \geq 1$, since $ct^{m-1} \notin uR(I)$. Thus, $P = Q \cap R = (uR(I) : ct^{m-1}) \cap R = (I^m : c)$, so $P \in \text{Ass } I^{m-1}/I^m \subseteq \text{Ass } R/I^m$.

ii) Since I is regular, (ii) follows from (1.15).

iii) This follows easily from (i), (ii), and the fact that (u, P, It) is the only irrelevant prime of R(I) which contains u and lies over P.

We now turn to the problem of determining when (u, P, It) is a prime divisor of uR(I).

(12.2) Lemma. Suppose that for some $n \geq 1$, $P = (I^n : b)$ for some $b \in R$. Suppose also that (u, P, It) is not minimal over $(u^n R(I) : bR(I))$. Then for all $k \geq 0$, $P = (I^{n+k} : bI^k)$.

Proof. We first note that $P = (I^n : b)$ easily implies that $(u^n R(I) : bR(I)) \subseteq (u, P, It)$. Let q be a prime minimal over $(u^n R(I) : bR(I))$ with $q \subset (u, P, It))$. Since $(u^n R(I) : bR(I)) \cap R = (I^n : b) = P$, and $(u, P, It) \cap R = P$, clearly $q \cap R = P$. Since $u \in q$ and $P \subseteq q$, we must have $It \nsubseteq q$. Let $c \in I$, with $ct \notin q$. For any $k \geq 0$, we have $(u^n R(I) : b(ct)^k R(I)) = ((u^n R(I) : bR(I)) : (ct)^k R(I)) \subseteq (q : (ct)^k R(I)) = q$. Contracting to R shows that $(I^{n+k} : bc^k) \subseteq P$. Therefore, $P = (I^n : b) \subseteq (I^{n+k} : bI^k) \subseteq (I^{n+k} : bc^k) \subseteq P$. Equality holds throughout.

(12.3) Theorem. The following are equivalent.

a) (u, P, It) is a prime divisor of uR(I).

b) $P \in$ Ass I^{n*}/I^n for some $n \geq 1$.

c) $P \in$ Ass $(I^{n+1} : I)/I^n$ for some $n \geq 1$.

d) $P \in$ Ass $((I^{n+1} : I) \cap I^{n-1})/I^n$ for some $n \geq 1$.

e) For some $n \geq 1$, there is an ideal J_n containing I^n with $P \in$ Ass J_n/I^n - Ass IJ_n/I^{n+1}. (Here, we allow the possibility that $J_n = R$.)

Proof. b) $\Rightarrow$ e) Suppose $P \in$ Ass I^{n*}/I^n for some $n \geq 1$. Now (11.1)(iv) shows that for large k, $I^{k*} = I^k$, so that $P \notin$ Ass I^{k*}/I^k. Thus, increasing n if necessary, we may assume that

$P \in$ Ass I^{n*}/I^n - Ass $(I^{n+1})^*/I^{n+1}$. Now $I(I^{n*}) \subseteq (I^{n+1})^*$, an easy exercise using (11.1)(i). Thus $P \in$ Ass I^{n*}/I^n - Ass $I(I^{n*})/I^{n+1}$. Therefore (e) holds, with $J_n = I^{n*}$.

e) $\Rightarrow$ a) Suppose (e) holds, and write $P = (I^n : b)$ with $b \in J_n$. Since

$bI \subseteq IJ_n$, and $P \notin$ Ass IJ_n/I^{n+1}, we must have $P \neq (I^{n+1} : bI)$. By (12.2), (u, P, It) must be minimal over $(u^n R(I) : bR(I))$. Thus (u, P, It) is a prime divisor of $u^n R(I)$, and hence of $uR(I)$, showing that (a) holds.

a) $\Rightarrow$ d) Suppose (u, P, It) is a prime divisor of $uR(I)$. Then (u, P, It) = $(uR(I) : bt^{n-1})$ for some homogeneous element bt^{n-1} in $R(I) - uR(I)$. Clearly, we must have $b \in I^{n-1} - I^n$, so $n \geq 1$. Intersecting with R, we see that $P = (I^n : b)$. Since $Itbt^{n-1} \subseteq uR(I)$, we have $b \in (I^{n+1} : I) \cap I^{n-1}$, so (d) holds.

d) $\Rightarrow$ c) $\Rightarrow$ b) are trivial (since (11.1)(v) shows $(I^{n+1} : I) \subseteq I^{n*}$).

(12.4) Corollary. i) If $P \in$ Ass R/I^n - Ass R/I^{n*} for some $n \geq 1$, then (u, P, It) is a prime divisor of $uR(I)$.

ii) If $P \in$ Ass R/I^n - Ass R/I^{n+1} for some $n \geq 1$, then (u, P, It) is a prime divisor of $uR(I)$.

iii) If $P \in$ Ass $\overline{I^n}/I^n$ - Ass $\overline{I^{n+1}}/I^{n+1}$ for some $n \geq 1$, then (u, P, It) is a prime divisor of $uR(I)$.

iv) If $P \in$ Ass I^{n-1}/I^n - Ass I^n/I^{n+1} for some $n \geq 1$, then (u, P, It) is a prime divisor of $uR(I)$.

Proof. For (i), the exact sequence $0 \to I^{n*}/I^n \to R/I^n \to R/I^{n*} \to 0$ shows that Ass $R/I^n \subseteq$ Ass $I^{n*}/I^n \cup$ Ass R/I^{n*}. If P is as in (i), we see that $P \in$ Ass I^{n*}/I^n. Thus (12.3)(b)$\Rightarrow$(a) implies the truth of (i). Now (ii), (iii), and (iv) all follow from (12.3)(e)$\Rightarrow$(a). For (ii), let $J_n = R$. For

(iii), let $J_n = \overline{I^n}$. For (iv), let $J_n = I^{n-1}$. In each case, we are given

$P \in$ Ass J_n/I^n - Ass J_{n+1}/I^{n+1}. Since $IJ_n \subseteq J_{n+1}$, we see that

$P \in$ Ass J_n/I^n - Ass IJ_n/I_{n+1}. By (12.3)(e)$\Rightarrow$(a), (ii), (iii), and (iv)

hold.

(12.5) Proposition. The following are equivalent.

a) $uR(I)$ has no irrelevant prime divisors.

b) $I^n = I^{n^*}$ for all $n \geq 1$.

c) $(I^{n+1} : I) = I^n$ for all $n \geq 1$.

d) $(I^{n+1} : I) \cap I^{n-1} = I^n$ for all $n \geq 1$.

e) There is a $k \geq 1$ with $(I^{n+k} : I^k) = I^n$ for all $n \geq 1$.

f) I has a strongly superficial element of some degree $k \geq 1$. (That is, for
some $k \geq 1$, there is a $b \in I^k$, with $(I^{n+k} : b) = I^n$ for all $n \geq 1$.)

Proof. (a)$\Leftrightarrow$(b)$\Leftrightarrow$(c)$\Leftrightarrow$(d) follows from (12.3)(a)$\Leftrightarrow$(b)$\Leftrightarrow$(c)$\Leftrightarrow$(d).

(c) $\Rightarrow$ (e) This is trivial.

(e) $\Rightarrow$ (b) Let (e) hold, and let $n \geq 1$. We claim that for any $m \geq 1$,
$(I^{n+mk} : I^{mk}) = I^n$. To see this, we induct on m, the key equation being
$(I^{n+mk} : I^{mk}) = ((I^{n+mk} : I^k) : I^{(m-1)k}) = (I^{n+(m-1)k} : I^{(m-1)k})$, the
last equality by (e). As (11.1)(v) shows that for large m we have
$(I^{n+mk} : I^{mk}) = I^{n^*}$, we see that (b) holds.

(f) $\Rightarrow$ (e) Let b be a strongly superficial element of degree $k \geq 1$ for I.
For any $n \geq 1$, $I^n \subseteq (I^{n+k} : I^k) \subseteq (I^{n+k} : b) = I^n$, and so (e) holds.

(a) $\Rightarrow$ (f) If $uR(I)$ has no irrelevant prime divisors, then a well known
graded version of the prime avoidance lemma allows us to find a

homogeneous element bt^k of $ItR(I)$ with bt^k a nonzero divisor modulo $uR(I)$. Of course $b \in I^k$. Since $bt^k \in ItR(I) - uR(I)$, we see that $k \geq 1$. For all $n \geq 1$, $(u^n R(I) : bt^k) = u^n R(I)$. Intersecting with R shows that $(I^{n+k} : b) = I^n$. Thus (f) holds.

(12.6) Remark: We wish to offer a second proof of $(12.5)(d) \Leftrightarrow (a)$. We easily see that $(uR(I) : ItR(I)) = \Sigma((I^{n+2} : I) \cap I^n)t^n$ over all integers n. Therefore, (d) holds If and only if this last expression equals $\Sigma I^{n+1}t^n = uR(I)$ if and only if $ItR(I)$ does not consist of zero divisors modulo $uR(I)$ if and only if (a) holds.

(12.7) Lemma. Let $n \geq 1$ be an integer.

a) $(u^n R(I) : <ItR(I)>) \cap R = I^{n*}$.

b) Suppose that $q_1 \cap ... \cap q_r \cap q_{r+1} \cap ... \cap q_s = u^n R(I)$ is a primary decomposition of $u^n R(I)$. Suppose that for $1 \leq i \leq s$, q_i is primary to Q_i. Suppose also that the ordering is such that $Q_1, ..., Q_r$ are relevant, while $Q_{r+1}, ..., Q_s$ are irrelevant. Then $(q_1 \cap ... \cap q_r) \cap R = I^{n*}$ (and $(q_1 \cap ... \cap q_s) \cap R = I^n$).

Proof. a) Note that for $k \geq 1$, $(u^n R(I) : (ItR(I))^k) \cap R = (I^{n+k} : I^k)$. By taking k large enough, the left hand side becomes $(u^n R(I) : <ItR(I)>) \cap R$ and, by (11.1)(v), the right hand side becomes I^{n*}, proving (a).

b) By (1.7)(a), $q_1 \cap ... \cap q_r$ is a primary decomposition of $u^n R(I) : <ItR(I)>$. Thus $(q_1 \cap ... \cap q_r) \cap R =$ $(u^n R(I) : <ItR(I)>) \cap R = I^{n*}$, using part (a). (Since $q_1 \cap ... \cap q_s = u^n R(I)$ and $u^n R(I) \cap R = I^n$, the parenthetical statement is trivial.)

(12.8) Remark. We wish to offer a second proof of $(12.3)(b) \Rightarrow (a)$, as we find the argument quite interesting. Suppose that $P \in \mathrm{Ass}\ I^{n*}/I^n$, and write $P = (I^n : b)$ with $b \in I^{n^*}$. Consider a primary decomposition of $u^n R(I)$ such as in (12.7)(b), (so that the Q_i are exactly the prime divisors of $uR(I)$). By (12.1)(i), we see that some of our primes Q_i must lie over P. To prove that (12.3)(a) holds, we must show that for some $i = r+1, ..., s$, Q_i lies over P, so that $Q_i = (u, P, It)$ is a prime divisor of $uR(I)$. Therefore, suppose to the contrary that $Q_1, ..., Q_w$ are the ones lying over

P, and that $w \le r$. Let $J = (\cap q_i) \cap R$, $1 \le i \le w$, and $L = (\cap q_i) \cap R$,

$w+1 \le i \le s$. By the parenthetical statement in (12.7)(b), $J \cap L = I^n$. Also,

(12.7)(b) shows that $I^{n^*} = (q_1 \cap ... \cap q_r) \cap R \subseteq (q_1 \cap ... \cap q_w) \cap R = J$.

Thus, $b \in I^{n^*} \subseteq J$. Now $P = (I^n : b) = (J \cap L : b) = (L : b)$ (since $b \in J$).

Thus P is a prime divisor of L. However, since $L = \cap(q_i \cap R)$,

$w+1 \le i \le s$, and since $q_i \cap R$ is primary to $Q_i \cap R$, clearly the prime

divisors of L come from among the $Q_i \cap R$ for $w+1 \le i \le s$, and P is not one of these. This is a contradiction.

(12.9) Proposition. If Q is an essential prime of $uR(I)$, then Q is relevant.

Proof. This follows from (1.16), our assumption that I is regular, and the fact that $E(uR(I)) = Q(uR(I))$ (using (3.14)). However, we now offer a second argument for showing this. We will let $\alpha(n) = \alpha(uR(I), ItR(I), n)$, as discussed in chapter 6. As we want to show

$ItR(I) \not\subseteq \cup\{Q \in E(uR(I))\}$, (6.2)(c),(a) shows that it will suffice to prove that $\alpha(n)$ is bounded above. It is easily seen that for $n \ge 1$ and $k \ge 1$, $(u^n R(I) : (ItR(I)^k) = \Sigma((I^{n+m+k} : I^k) \cap I^m)t^m$ over all integers m. By (11.1)(v), for all $h \ge 1$, there is a $k = k(h) \ge 1$ with $(I^{h+k} : I^k) = I^{h^*}$. It follows from (11.1)(iv) that for large h, we may take $k(h) = 1$

(since $I^h \subseteq (I^{h+1} : I) \subseteq I^{h^*} = I^h$). Therefore, having to really only worry about finitely many small h, we see that there is a fixed $k \ge 1$, independent of h, such that for all $h \ge 1$, $(I^{h+k} : I^k) = I^{h^*}$. For such a k, $(u^n R(I) : (ItR(I)^k) = \Sigma((I^{n+m})^* \cap I^m)t^m$ over all integers m (with $(I^{n+m})^* = R$ when $n + m \le 0$). Therefore, increasing k does not change $(u^n R(I) : (ItR(I)^k)$, which shows that $\alpha(n) \le k$. This is true of all $n \ge 1$, and so we are done.

(12.10) Example. As (11.3) shows, the sequence Ass R/I^n, $n \geq 1$, need not

be increasing. It is harder to see that the sequence Ass $R/I^n \cap A^*(I)$ also

need not be increasing, particularly if we insist that I be regular. We

reproduce an example due to R. C. Cowsik. Let F be a field, and

X, Y, T, Z be indeterminates. Let $A = F[X, Y, T, Z]/(X^m + Y^p - TZ^q)$,

(for some large m, p, q; m, p >> q). The domain A contains

$R = F[X^4, X^5, X^{11}, Y^4, Y^5, Y^{11}, Z^4, Z^5, Z^{11}, T, TX^6Y^6Z^6]$.

Let $P = (X^4, X^5, X^{11}, Y^4, Y^5, Y^{11}, Z^4, Z^5, Z^{11}, T, TX^6Y^6Z^6)$, and

$Q = (X^4, X^5, X^{11}, Y^4, Y^5, Y^{11}, Z^4, Z^5, Z^{11}, TX^6Y^6Z^6)$. Now Q is prime,

and $X^{11}Y^{11}Z^{11} \in Q^{(4)} - Q^4$, while $Q^{(5)} = Q^5$. Also, $Q^{(n)} \neq Q^n$ for all

large n. It follows that $P \in$ Ass R/Q^4 and $P \in$ Ass R/Q^n for all large n

(so $P \in A^*(Q)$), but $P \notin$ Ass R/Q^5. Since $P \in$ Ass R/Q^4 - Ass R/Q^5, by

(12.4)(i), (u, P, Qt) is a prime divisor of uR(Q). Since $P \in A^*(Q)$, by

(12.1)(iii), (u, P, Qt) is not the only prime divisor of uR(Q) which

intersects R at P.

This material reworks results in [MR4, section 4].

13   GENERALIZATIONS TO MANY IDEALS

Notation. Let $I_1, \ldots, I_g$ be finitely many ideals in the Noetherian ring R, and let $\mathbf{N_g}$ be the set of all g-tuples of nonnegative integers.

For $\mathbf{n} = (n_1, \ldots, n_g) \in \mathbf{N_g}$, let $I^{\mathbf{n}} = I_1^{n_1} \ldots I_g^{n_g}$. For $1 \leq i \leq g$, $\mathbf{n}(i)$ will mean n_i, the i-th component of $\mathbf{n}$. If $\mathbf{n}$ and $\mathbf{m}$ are in $\mathbf{N_g}$, we will write $\mathbf{n} \geq \mathbf{m}$ (respectively, $\mathbf{n} > \mathbf{m}$) if for each $1 \leq i \leq g$, $\mathbf{n}(i) \geq \mathbf{m}(i)$ (respectively, $\mathbf{n}(i) > \mathbf{m}(i)$). For $h \geq 0$ an integer, we will define $h\mathbf{n}$ and $\mathbf{n} \pm \mathbf{m}$ in the usual componentwise manner, ($\mathbf{n} - \mathbf{m}$ being defined only when $\mathbf{n} \geq \mathbf{m}$). Furthermore, J will be another ideal of R.

The purpose of this chapter is to show that if $I_1, \ldots, I_g$ are regular ideals, then for all infinite sequences $\mathbf{n}_1 \leq \mathbf{n}_2 \leq \mathbf{n}_3 \leq \ldots$ in $\mathbf{N_g}$, the sequence Ass $R/I^{\mathbf{n}_1}$, Ass $R/I^{\mathbf{n}_2}$, Ass $R/I^{\mathbf{n}_3}$, $\cdots$ eventually stabilizes, as do the sequences Ass $R/ \overline{I^{\mathbf{n}_1}}$, Ass $R/ \overline{I^{\mathbf{n}_2}}$, Ass $R/ \overline{I^{\mathbf{n}_3}}$, $\cdots$ and Ass $R/(I^{\mathbf{n}_1})^*$, Ass $R/(I^{\mathbf{n}_2})^*$, Ass $R/(I^{\mathbf{n}_3})^*$, $\cdots$. The results in this chapter are taken from [KMR].

(13.1) Lemma. $\cup\{\text{Ass } R/I^{\mathbf{n}} \mid \mathbf{n} \in \mathbf{N_g}\}$ is finite.

Proof. Let $\mathbf{R} = R[I_1 t_1, \ldots, I_g t_g, t_1^{-1}, \ldots, t_g^{-1}]$ be the Rees ring of R with respect to $I_1, \ldots, I_g$. For $\mathbf{n} \in \mathbf{N_g}$, $(t_1^{-\mathbf{n}(1)} \ldots t_g^{-\mathbf{n}(g)})\mathbf{R} \cap R = I^{\mathbf{n}}$. Thus any prime divisor P of $I^{\mathbf{n}}$ lifts to a prime divisor $\mathbf{P}$ of $(t_1^{-\mathbf{n}(1)} \ldots t_g^{-\mathbf{n}(g)})\mathbf{R}$. For some $1 \leq i \leq g$, $t_i^{-1} \in \mathbf{P}$, and since each t_i^{-1} is regular in $\mathbf{R}$, $\mathbf{P}$ is a prime divisor of $t_i^{-1}\mathbf{R}$. Therefore, every prime in $\cup\{\text{Ass } R/I^{\mathbf{n}} \mid \mathbf{n} \in \mathbf{N_g}\}$ lifts to a prime in $\cup\{\text{Ass } \mathbf{R}/t_i^{-1}\mathbf{R} \mid 1 \leq i \leq g\}$. As this last set is finite, the result follows.

We need a version of the Artin-Rees lemma, (supplied by D. Katz).

(13.2) Proposition. a) If A is a finitely generated R-module, and if B is a submodule of A, then there is an integer $k \geq 0$, such that for all integers $m \geq k$, and all $\mathbf{n} \in \mathbf{N}_g$, $J^m I^{\mathbf{n}} A \cap B = J^{m-k}(J^k I^{\mathbf{n}} A \cap B)$.

b) There is an integer $h \geq 0$ such that for all integers $m \geq h$, all integers $r \geq 0$, and all $\mathbf{n} \in \mathbf{N}_g$, $(J^{m+r} I^{\mathbf{n}} : J^r) \cap J^h I^{\mathbf{n}} = J^m I^{\mathbf{n}}$.

c) If J is regular, then there is an integer $k \geq 0$ such that for all integers $m \geq k$, all integers $r \geq 0$, and all $\mathbf{n} \in \mathbf{N}_g$, $(J^{m+r} I^{\mathbf{n}} : J^r) = J^m I^{\mathbf{n}}$.

Proof. a) For $s = (s_0, s_1, ..., s_g) \in \mathbf{N}_{g+1}$, and $t_0, t_1, ..., t_g$ indeterminates, let $H^s = J^{s_0} I_1^{s_1} ... I_g^{s_g}$, and $t^s = t_0^{s_0} t_1^{s_1} ... t_g^{s_g}$. Let $\mathbf{R} = R[Jt_0, I_1 t_1, ..., I_g t_g]$ be the (restricted) Rees ring of R with respect to $J_0, I_1, ..., I_g$. Let $\mathbf{A}$ consist of all finite sums of the form $\Sigma a_s t^s$ with $s \in \mathbf{N}_{g+1}$ and $a_s \in H^s A$. It is easily seen that $\mathbf{A}$ is an $\mathbf{N}_g$ graded finitely generated $\mathbf{R}$-module. Let $\mathbf{B}$ be the $\mathbf{R}$-submodule of $\mathbf{A}$ consisting of those $\Sigma a_s t^s$ such that $a_s \in H^s A \cap B$. Now $\mathbf{B}$ has a finite set of homogeneous generators. Let k be the maximum value of any exponent of t_0 which appears among those generators. It is easily seen that (a) holds for this k.

b) Let $\mathbf{D} = (J\mathbf{R} : Jt_0\mathbf{R})$, a homogeneous ideal of $\mathbf{R}$. Consider a finite set of homogeneous generators for $\mathbf{D}$, and let h be one more than the highest power of t_0 occurring within those generators. Suppose that $m \geq h$, and let $x \in (J^{m+1} I^{\mathbf{n}} : J) \cap J^h I^{\mathbf{n}}$ for some $\mathbf{n} = (n_1, ..., n_g) \in \mathbf{N}_g$. We claim that $x \in J^m I^{\mathbf{n}}$. If $m = h$, this is trivial. Thus, take $m > h$. Let $s = (h, n_1, ..., n_g) \in \mathbf{N}_{g+1}$. Since $x \in J^h I^{\mathbf{n}}$, we have $xt^s \in \mathbf{R}$. Since $x \in (J^{m+1} I^{\mathbf{n}} : J)$, and since $m > h$, we see that $xt^s \in \mathbf{D}$. Considering the above mentioned generators for $\mathbf{D}$, and recalling the definition of h, we see that xt^s can be written as a linear combination of some of those generators, with coefficients coming from $Jt_0\mathbf{R}$. However, anything in $Jt_0\mathbf{R}$ times anything in $\mathbf{D}$ is in $J\mathbf{R}$. Thus $xt^s \in J\mathbf{R}$. This implies that $x \in J^{h+1} I^{\mathbf{n}}$. If $m = h + 1$, the claim is proved. If $m > h + 1$, we may iterate the argument, finally concluding that $x \in J^m I^{\mathbf{n}}$. This proves the claim. The claim obviously implies the truth of (b) when $r = 1$. The case $r > 1$ follows by an easy induction, and the case $r = 0$ is trivial. This proves (b).

c) Assume that J is regular. Using [Kp, Theorem 124], there is a set of regular elements $y_1, \ldots, y_s$ which generate J. Let

$A = Ry_1^{-1} \oplus \ldots \oplus Ry_s^{-1}$. Let $B = \{(ay_1/y_1, \ldots, ay_s/y_s) \mid a \in R\}$, an

R-submodule of A. If K is any ideal of R, we see that $KA \cap B =$

$\{(a, \ldots, a) \mid a \in (K:J)\}$. Let k be as in part (a) applied to this A and B.

Then for all $m \geq k$ and $\mathbf{n} \in \mathbf{N_g}$, $J^{m+1}I^{\mathbf{n}}A \cap B = J^{m+1-k}(J^kI^{\mathbf{n}}A \cap B)$. It

follows that $(J^{m+1}I^{\mathbf{n}} : J) = J^{(m-k)+1}(J^kI^{\mathbf{n}} : J) \subseteq J^kI^{\mathbf{n}}$. It does no harm to increase the value of k, and so we assume that k is at least as large as the h in part (b). Now (b) shows that (c) holds when $r = 1$. The rest of (c) follows easily.

(13.3) Corollary. a) Let $H_1, \ldots, H_h$ be regular ideals of R. Let $L_1, \ldots, L_d$ be any ideals of R. (Possibly $d = 0$.) Then there is a $\mathbf{k} \in \mathbf{N_h}$ such that for all s and t in $\mathbf{N_h}$ with $s \geq k$ and all $\mathbf{r} \in \mathbf{N_d}$, $(H^{s+t}L^{\mathbf{r}} : H^t) = H^sL^{\mathbf{r}}$.

b) Let $I_1, \ldots I_g$ be regular. Then there is a $\mathbf{k} \in \mathbf{N_g}$ such that for all $\mathbf{n} \geq \mathbf{k}$, $(I^{\mathbf{n}})^* = I^{\mathbf{n}}$. (Recall that $(I^{\mathbf{n}})^*$ was defined in chapter 11.)

Proof. a) Fix an i, $1 \leq i \leq h$. Let $g = h + d - 1$. Let $I_1, \ldots, I_g$ be $H_1, \ldots, H_{i-1}$, $H_{i+1}, \ldots H_h, L_1, \ldots, L_d$, and let J be H_i. Let k_i be the k of (13.2)(c) for this situation. For each $1 \leq i \leq h$, find this k_i, and let $\mathbf{k} = (k_1, \ldots, k_h)$. The present result now follows from repeated uses of (13.2)(c) and the fact that for ideals K_1, K_2, and K_3, $(K_1 : K_2K_3) = ((K_1 : K_2) : K_3)$.

b) We first note that part (a) applied to the case $d = 0$, $h = g$, and $H_i = I_i$, for $1 \leq i \leq g$, shows there is a $\mathbf{k} \in \mathbf{N_g}$ such that for all $\mathbf{n}$ and $\mathbf{m}$ in $\mathbf{N_g}$ with $\mathbf{n} \geq \mathbf{k}$, $(I^{\mathbf{n}+\mathbf{m}} : I^{\mathbf{m}}) = I^{\mathbf{n}}$. Now the definition of $(I^{\mathbf{n}})^*$ shows that for some integer $h \geq 0$, $(I^{\mathbf{n}})^* = ((I^{\mathbf{n}})^{h+1} : (I^{\mathbf{n}})^h)$. Letting $\mathbf{m} = h\mathbf{n}$, we have $(I^{\mathbf{n}})^* = (I^{\mathbf{n}+\mathbf{m}} : I^{\mathbf{m}}) = I^{\mathbf{n}}$, for all $\mathbf{n} \geq \mathbf{k}$.

We now reach our first goal.

(13.4) Theorem. Let $I_1, \ldots, I_g$ be regular. If $\mathbf{n_1} \leq \mathbf{n_2} \leq \mathbf{n_3} \leq \ldots$ is a sequence in $\mathbf{N_g}$, the sequence Ass $R/I^{\mathbf{n}1}$, Ass $R/I^{\mathbf{n}2}$, Ass $R/I^{\mathbf{n}3}$, … eventually stabilizes. Furthermore, there is a $\mathbf{d} \in \mathbf{N_g}$ such that Ass $R/I^{\mathbf{n}}$ is independent of $\mathbf{n}$ for $\mathbf{n} \geq \mathbf{d}$.

Proof. By (13.1), to prove the first part of the result it will suffice to show that for large j, Ass $R/I^{n}j \subseteq$ Ass $R/I^{n}j+1$. We may assume that for $1 \leq i \leq h$, that $\{n_j(i) \mid j \geq 1\}$ is infinite, while for $h + 1 \leq i \leq g$, $\{n_j(i) \mid j \geq 1\}$ is finite. Ignoring small values of j, we may assume that $n_j = (n_j(1), \ldots n_j(h), r_{h+1}, \ldots r_g)$, with $r_{h+1}, \ldots, r_g$ constants and $n_j(1), \ldots, n_j(h)$ arbitrarily large. Let $d = g - h$, let $H_1, \ldots, H_h$ be $I_1, \ldots, I_h$, and let $L_1, \ldots, L_d$ be $I_{h+1}, \ldots, I_g$. Also let $s_j = (n_j(1), \ldots n_j(h))$ $\in N_h$, and $r = (r_{h+1}, \ldots, r_g) \in N_d$. Thus, $I^{n}j = H^{s}jL^{r}$. Suppose that $P \in$ Ass $R/I^{n}j$. For some $x \in R$, write $P = (I^{n}j : x) = (H^{s}jL^{r} : x)$. Let $t_j = s_{j+1} - s_j$. Since we may assume that $s_j \geq k$, with k as in (13.3)(a), we have $(I^{n}j+1 : xH^{t}j) = (H^{(s_j+t_j)}L^{r} : xH^{t}j) = ((H^{(s_j+t_j)}L^{r} : H^{t}j) : x) =$ (by (13.3)(a)) $(H^{s}jL^{r} : x) = (I^{n}j : x) = P$. This shows that $P \in$ Ass $R/I^{n}j+1$. The first part of the result is now proven.

For the second part, as seen in the proof of (13.3)(b), there is a $k \in N_g$ such that for any n and m in N_g with $n \geq k$, $(I^{n+m} : I^{m}) = I^{n}$. By an argument similar to that in the previous paragraph, it is seen that for $s \geq t \geq k$, Ass $R/I^t \subseteq$ Ass R/I^s. Now applying the first part of the result to $n_j = (j, \ldots, j), j \geq 1$, we see that we can find j large enough that Ass $R/I^{n}j =$ Ass $R/I^{hn}j$ for all integers $h \geq 1$. We can at the same time take $n_j \geq k$. Let d equal that n_j, so that $d \geq k$ and Ass $R/I^d =$ Ass R/I^{hd} for all $h \geq 1$. Suppose that $n \geq d$. Pick h such that $hd \geq n \geq d$. Since $d \geq k$, we have Ass $R/I^d \subseteq$ Ass $R/I^n \subseteq$ Ass $R/I^{hd} =$ Ass R/I^d. Thus Ass $R/I^n =$ Ass R/I^d, proving the second part of the result.

(13.5) Questions. 1) Can we drop the assumption that $I_1, \ldots, I_g$ be regular from (13.4)?

2) Can we strengthen (13.4) to say there is an integer $h \geq 1$ such that for all $n_1 \leq n_2 \leq n_3 \leq \ldots$ in N_g, Ass $R/I^{n}j$ is stable for $j \geq h$?

In order to reach the remaining goals of this chapter, we will use Δ-closures, a device introduced by Ratliff in [R6].

Notation. In the rest of this chapter, Δ will represent a nonempty, multiplicatively closed set of nonzero ideals in R.

Definition. Let H be an ideal in R. Let H_Δ be the unique maximal element of the set $\{(HK : K) \mid K \in \Delta\}$. (Since for K and L in Δ, $(HKL : KL)$ contains both $(HK : K)$ and $(HL : L)$, and since R is Noetherian, we see that the set under consideration does in fact have a unique maximal element.)

(13.6) Lemma. Suppose that $\{I^n \mid n \in N_g\} \subseteq \Delta$. Then for $n \geq k$ elements of N_g, Ass $R/(I^k)_\Delta \subseteq$ Ass $R/(I^n)_\Delta$.

Proof. Let $P = ((I^k)_\Delta : x) \in$ Ass $R/(I^k)_\Delta$, with $x \in R$. For some $K \in \Delta$, $(I^k)_\Delta = (I^k K : K)$. Now $I^{n-k}(I^k)_\Delta = I^{n-k}(I^k K : K) \subseteq (I^n K : K) \subseteq (I^n)_\Delta$. Thus, $P = ((I^k)_\Delta : x) \subseteq (I^{n-k}(I^k)_\Delta : xI^{n-k}) \subseteq ((I^n)_\Delta : xI^{n-k})$. However, we claim this last ideal is contained in P. Let y belong to this last ideal. Then $yx \in ((I^n)_\Delta : I^{n-k})$. For some $L \in \Delta$, $(I^n)_\Delta = (I^n L : L)$, so $yx \in ((I^n L : L) : I^{n-k}) = (I^k I^{n-k} L : I^{n-k} L) \subseteq (I^k)_\Delta$, since $I^{n-k} L \in \Delta$. Thus $y \in ((I^k)_\Delta : x) = P$, proving the claim. Therefore, $P = ((I^n)_\Delta : xI^{n-k})$, showing that $P \in$ Ass $R/(I^n)_\Delta$.

(13.7) Proposition. Suppose that Δ is such that $\{I^n \mid n \in N_g\} \subseteq \Delta$, and $\cup\{$Ass $R/(I^m)_\Delta \mid m \in N_g\}$ is finite. Then for any sequence $n_1 \leq n_2 \leq n_3 \leq \dots$ in N_g, the sequence Ass $R/(I^{n_1})_\Delta \subseteq$ Ass $R/(I^{n_2})_\Delta \subseteq$ Ass $R/(I^{n_3})_\Delta \subseteq \dots$ eventually stabilizes. Furthermore, there is a $d \in N_g$ such that Ass $R/(I^n)_\Delta$ is independent of n, for all $n \geq d$.

Proof. By (13.6), the sequence of Ass $R/(I^{n_j})_\Delta$ is an increasing sequence. The first conclusion obviously follows from the assumption that $\cup\{$Ass $R/(I^m)_\Delta \mid m \in N_g\}$ is finite. For the second conclusion, choose d essentially the same way d was chosen in the proof of (13.4).

We will now present two choices of Δ for which the hypothesis of (13.7) is satisfied, and at the same time, $(I^n)_\Delta$ is "of interest". We begin with the case that Δ consists of all regular ideals. In this case, we will see that $(J)_\Delta = \overline{J}$, whenever J is regular.

(13.8) Theorem. Let $I_1, ..., I_g$ be regular ideals. Then for any sequence $n_1 \le n_2 \le n_3 \le ...$ in N_g, the sequence Ass R/ $\overline{I^{n_1}} \subseteq$ Ass R/ $\overline{I^{n_2}} \subseteq$ Ass R/ $\overline{I^{n_3}} \subseteq ...$ eventually stabilizes. Furthermore, there is a $d \in N_g$ such that Ass R/ $\overline{I^n}$ is independent of n, for all $n \ge d$.

Proof. Let Δ be the set of all regular ideals. We claim that for any regular ideal J, $J_\Delta = \overline{J}$. Say $J_\Delta = (JK : K)$ for $K \in \Delta$, and suppose that $K = (k_1, ..., k_r)R$. Then for $x \in J_\Delta$ and $1 \le i \le r$, we have $xk_i = \Sigma a_{ij}k_j$, $1 \le j \le r$, with $a_{ij} \in J$. Since K is regular, a standard determinant argument now shows $x \in \overline{J}$. (This much of the argument shows that if every ideal in Δ is regular, then for any ideal J, $J_\Delta \subseteq \overline{J}$.) Now let $x \in \overline{J}$. Then $J(J, x)^n = (J, x)^{n+1}$ for some $n \ge 0$. Thus $x(J, x)^n \subseteq J(J, x)^n$, so (with K as before), $xK(J, x)^n \subseteq JK(J, x)^n$. However, since $K(J, x)^n \in \Delta$, it follows that $J_\Delta = (JK(J, x)^n : K(J, x)^n)$. Thus we see that $x \in J_\Delta$, so that $\overline{J} \subseteq J_\Delta$, proving our claim Now for any $m \in N_g$, Ass $R/(I^m)_\Delta$ = Ass R/ $\overline{I^m} \subseteq \overline{A}^*(I^m) \subseteq A^*(I^m) \subseteq$ $\cup\{$Ass R/I^n | $n \in N_g\}$, using (1.2), (1.1)(d), and the definition of $A^*(I^m)$. This, together with (13.1), shows that $\cup\{$Ass $R/(I^m)_\Delta$ | $m \in N_g\}$ is finite. Since $\{I^n$ | $n \in N_g\} \subseteq \Delta$, the hypothesis of (13.7) is satisfied. That result, together with the above claim, implies the present result.

In our next application of (13.7), we again take $I_1, ..., I_g$ to be regular, but take $\Delta = \{I^m$ | $m \in N_g\}$. We will see that $(I^n)_\Delta = (I^n)^*$ whenever $n \ge (1, ..., 1)$.

(13.9) Lemma. Let $I_1, ..., I_g$ be regular ideals.

a) Suppose $\mathbf{n}$ and $\mathbf{m}$ are in $\mathbf{N}_g$ with $\mathbf{n} \geq (1, ..., 1)$. Let k be an integer with $k\mathbf{n} \geq \mathbf{m}$. Then $(I^{\mathbf{n+m}} : I^{\mathbf{m}}) \subseteq ((I^{\mathbf{n}})^{k+1} : (I^{\mathbf{n}})^k) \subseteq (I^{\mathbf{n}})^*$.
(Note: Clearly such k exist.)

b) If $\Delta = \{I^{\mathbf{m}} \mid \mathbf{m} \in \mathbf{N}_g\}$ then for $\mathbf{n} \geq (1, ..., 1)$, $(I^{\mathbf{n}})^* = (I^{\mathbf{n}})_\Delta$.

Proof. For (a), let $x \in (I^{\mathbf{n+m}} : I^{\mathbf{m}})$. Since $k\mathbf{n} - \mathbf{m} \in \mathbf{N}_g$, we may write $(I^{\mathbf{n}})^k = I^{\mathbf{m}}I^{k\mathbf{n-m}}$. Thus $x(I^{\mathbf{n}})^k = xI^{\mathbf{m}}I^{k\mathbf{n-m}} \subseteq I^{\mathbf{n+m}}I^{k\mathbf{n-m}} = (I^{\mathbf{n}})^{k+1}$. This gives the first containment of (a). The second containment is by the definition of $(I^{\mathbf{n}})^*$. For (b), suppose that $\Delta = \{I^{\mathbf{m}} \mid \mathbf{m} \in \mathbf{N}_g\}$ and $\mathbf{n} \geq (1, ..., 1)$. Then for large integers h, $(I^{\mathbf{n}})^* = ((I^{\mathbf{n}})^{h+1} : (I^{\mathbf{n}})^h) = (I^{\mathbf{n}}I^{h\mathbf{n}} : I^{h\mathbf{n}}) \subseteq (I^{\mathbf{n}})_\Delta$, since $I^{h\mathbf{n}} \in \Delta$. For the reverse inclusion, there is an $\mathbf{m} \in \mathbf{N}_g$ with $(I^{\mathbf{n}})_\Delta = (I^{\mathbf{n+m}} : I^{\mathbf{m}})$. By (a), $(I^{\mathbf{n}})_\Delta \subseteq (I^{\mathbf{n}})^*$.

(13.10) Example. In (13.9), we need $\mathbf{n} \geq (1, ..., 1)$. By [RR, (3.4) and (4.2)], there exist regular ideals I_1 and I_2 with I_1^* properly contained in $(I_1I_2 : I_2)$. Let $\mathbf{n} = (1, 0)$ and $\mathbf{m} = (0, 1)$. Then $(I^{\mathbf{n+m}} : I^{\mathbf{m}}) = (I_1I_2 : I_2)$ is not contained in $(I^{\mathbf{n}})^* = I_1^*$.

(13.11) Theorem. Let $I_1, ..., I_g$ be regular ideals. Then for any sequence $\mathbf{n}_1 \leq \mathbf{n}_2 \leq \mathbf{n}_3 \leq ...$ in $\mathbf{N}_g$, the sequence Ass $R/(I^{\mathbf{n}1})^*$, Ass $R/(I^{\mathbf{n}2})^*$, Ass $R/(I^{\mathbf{n}3})^*$, ... eventually stabilizes. Furthermore, there is a $\mathbf{d} \in \mathbf{N}_g$ such that Ass $R/(I^{\mathbf{n}})^*$ is independent of $\mathbf{n}$, for all $\mathbf{n} \geq \mathbf{d}$.

(Remark: If $\mathbf{n}_1 \geq (1, ..., 1)$, then the sequence of Ass $R/(I^{\mathbf{n}i})^*$ is increasing, by (13.9)(b) and (13.6). However, in the example (13.10), we cannot deduce that Ass $R/(I^{\mathbf{n}})^* \subseteq$ Ass $R/(I^{\mathbf{n+m}})^*$.)

Proof. Let $\Delta = \{I^{\mathbf{n}} \mid \mathbf{n} \in \mathbf{N}_g\}$. For some $\mathbf{m} \in \mathbf{N}_g$, let $P \in$ Ass $R/(I^{\mathbf{m}})_\Delta$. For some $x \in R$ and $I^{\mathbf{n}} \in \Delta$, we may write $P = ((I^{\mathbf{m}})_\Delta : x) = ((I^{\mathbf{m}}I^{\mathbf{n}} : I^{\mathbf{n}}) : x) = (I^{\mathbf{m+n}} : xI^{\mathbf{n}})$. This shows that $P \in$ Ass $R/I^{\mathbf{m+n}}$. By (13.1), we now see that $\cup\{$Ass $R/(I^{\mathbf{m}})_\Delta \mid \mathbf{m} \in \mathbf{N}_g\}$ is finite. Therefore, we see that the hypothesis of (13.7) holds. Now we can see that the first part of the present result holds for any sequence $n_1 \leq n_2 \leq n_3 \leq \ldots$ such that for large j, $n_j \geq (1, \ldots, 1)$, since then (13.9)(b) shows that Ass $R/(I^{n_j})^* =$ Ass $R/(I^{n_j})_\Delta$, and so (13.7) shows that the portion of the sequence from n_j onward eventually stabilizes. On the other hand, if there is no j with $n_j \geq (1, \ldots, 1)$, then for some $1 \leq i \leq g$, we have $n_j(i)$, the i-th component of n_j, equaling zero for all j. However, for such an i, this says that I_i is not involved in our considerations. Thus we can ignore all such i and I_i, and lower the value of g appropriately. Thus the case just mentioned, applied to this smaller value of g, shows that the first part of the result always holds. For the second part of the result, take $\mathbf{d}$ as in (13.7). It does no harm to increase $\mathbf{d}$, and so also assume that $\mathbf{d} \geq (1, \ldots, 1)$, so that for $\mathbf{n} \geq \mathbf{d}$, $(I^{\mathbf{n}})^* = (I^{\mathbf{n}})_\Delta$ by (13.9)(b). Now the second conclusion of (13.7) gives the second conclusion of the present result.

(13.12) Remark. The second conclusion of (13.11) and the second conclusion of (13.4) contain the same information, (via (13.3)(b)). However, the first parts of those two results contain somewhat different information.

(13.13). Question: We have shown two choices of Δ which satisfy the hypothesis of (13.7). Are there other choices of Δ which satisfy it? (Remark: In the two cases we considered, both had

$\cup\{$Ass $R/(I^{\mathbf{m}})_\Delta \mid \mathbf{m} \in \mathbf{N}_g\} \subseteq \cup\{$Ass $R/I^{\mathbf{n}} \mid \mathbf{n} \in \mathbf{N}_g\}$. It is interesting to note that two cases required very different arguments to show that containment.)

Notation. Fix a choice of Δ. Let $\mathbf{R} = \Sigma I^{\mathbf{n}} t^{\mathbf{n}}$ and $\mathbf{R}_\Delta = \Sigma (I^{\mathbf{n}})_\Delta t^{\mathbf{n}}$ over $\mathbf{n} \in \mathbf{N}_g$. (We easily see that $(I^{\mathbf{n}})_\Delta (I^{\mathbf{m}})_\Delta \subseteq (I^{\mathbf{n}+\mathbf{m}})_\Delta$, so that $\mathbf{R}_\Delta$ is a ring and an $\mathbf{R}$-module.)

In the remainder of this chapter, we will investigate the nice consequences of assuming that $\mathbf{R}_\Delta$ is a finite $\mathbf{R}$-module. We first show that this assumption is sometimes valid.

(13.14) Theorem. Let R be a locally analytically unramified ring whose integral closure is a finite R-module. Let $I_1, ..., I_g$ be regular, and suppose that all the ideals in Δ are regular. Then R_Δ is a finite R-module.

Proof. The parenthetical comment in the proof of (13.8) shows that $(I^n)_\Delta \subseteq \overline{I^n}$. Thus $R \subseteq R_\Delta \subseteq \Sigma \overline{I^n} t^n$ over $n \in N_g$. Therefore, R_Δ is contained in the integral closure of R. However, the hypothesis on R implies that the integral closure of R is a finite R-module. (This is proved using arguments similar to those in [N3. (35.3)]. See [KR2, Lemma 1].) The result follows.

(13.15) Proposition. Suppose R_Δ is a finite R-module. Then

a) $\cup\{$Ass $R/(I^n)_\Delta \mid n \in N_g\}$ is a finite set.

b) There is a $K \in \Delta$ such that for all $n \in N_g$, $(I^n)_\Delta = (IK : K)$.

c) There is a $b \in N_g$ such that if $n \geq m \geq b$, then $(I^n)_\Delta = I^{n-m}(I^m)_\Delta$.

d) Suppose $\{I^n \mid n \in N_g\} \subseteq \Delta$. Then for any $n_1 \leq n_2 \leq n_3 \leq ...$ in N_g, the sequence Ass $R/(I^{n_1})_\Delta \subseteq$ Ass $R/(I^{n_2})_\Delta \subseteq$ Ass $R/(I^{n_3})_\Delta \subseteq ...$ eventually stabilizes. Also, there is a $d \in N_g$ such that Ass $R/(I^n)_\Delta$ is independent of n for all $n \geq d$.

e) Suppose that $I_1, ..., I_g$ are regular, and $\{I^n \mid n \in N_g\} \subseteq \Delta$. Then there is an integer $k \geq 0$ such that for all $n \in N_g$, $(I^n)^* = ((I^n)^{k+1} : (I^n)^k)$.

Proof. a) Let $S_\Delta = R_\Delta[t_1^{-1}, ..., t_g^{-1}]$. Since R_Δ is a finite R-module, S_Δ is a Noetherian ring. Since for $n \in N_g$, $t^{-n}S_\Delta \cap R = (I^n)_\Delta$, any prime in Ass $R/(I^n)_\Delta$ lifts to a prime in Ass $S_\Delta/t^{-n}S_\Delta \subseteq \cup\{$Ass $S_\Delta/t_i^{-1}S_\Delta \mid 1 \leq i \leq g\}$. As this last set is finite, (a) holds.

b) Since R_Δ is a finite R-module, there are finitely many $m_1, ..., m_r$ in N_g with $R_\Delta = R((I^{m_1})_\Delta t^{m_1}) + ... + R((I^{m_r})_\Delta t^{m_r})$. For $1 \leq j \leq r$, let $K_j \in \Delta$ with $(I^{m_j})_\Delta = (I^{m_j}K_j : K_j)$. Let $K = K_1 ... K_r$. Clearly

$(I^m j)_\Delta = (I^m jK : K)$ for $1 \le j \le r$. Now let $\mathbf{T}$ be the $\mathbf{R}$-submodule of $\mathbf{R}_\Delta$ having the form $\mathbf{T} = \Sigma(I^n K : K)t^n$ over all $\mathbf{n} \in \mathbf{N}_g$. (Since $I^m(I^n K : K) \subseteq (I^{n+m}K : K)$, this is a submodule.) Now $\mathbf{T}$ contains $(I^m jK : K)t^{m_j} = (I^m j)_\Delta t^{m_j}$ for $1 \le j \le r$, and as these generate $\mathbf{R}_\Delta$ over $\mathbf{R}$, we see that $\mathbf{T} = \mathbf{R}_\Delta$. It follows that for all $\mathbf{n} \in \mathbf{N}_g$, $(I^n)_\Delta = (I^n K : K)$, proving (b).

(c) With $m_1, ..., m_r$ as in the proof of (b), pick $\mathbf{b}$ such that $\mathbf{b} \ge m_j$ for $1 \le j \le r$. Suppose that $\mathbf{n} \ge \mathbf{m} \ge \mathbf{b}$. Since $\mathbf{R}_\Delta = \mathbf{R}((I^{m_1})_\Delta t^{m_1}) + ... + \mathbf{R}((I^{m_r})_\Delta t^{m_r})$, by considering the t^n-th component of $\mathbf{R}_\Delta$, we see that $(I^n)_\Delta = \Sigma I^{(n-m_j)}(I^{m_j})_\Delta$ over $1 \le j \le r$ (since for each such j, $\mathbf{n} \ge \mathbf{b} \ge m_j$). Similarly, $(I^m)_\Delta = \Sigma I^{(m-m_j)}(I^{m_j})_\Delta$ over $1 \le j \le r$. Thus, $(I^n)_\Delta = I^{n-m}(I^m)_\Delta$, proving (c).

d) This follows from (13.7) and part (a).

e) Let $\mathbf{R}_* = \Sigma(I^n)^* t^n$, and for $k \ge 0$, let $\mathbf{R}_k = \Sigma((I^n)^{k+1} : (I^n)^k)t^n$, both sums over all $\mathbf{n} \in \mathbf{N}_g$. We claim that $\mathbf{R}_* = \cup \mathbf{R}_k$, $k \ge 0$. Surely $\mathbf{R}_k \subseteq \mathbf{R}_*$. Now for any $\mathbf{n} \in \mathbf{N}_g$, for large enough k we have $(I^n)^* = ((I^n)^{k+1} : (I^n)^k)$, so that $(I^n)^* t^n \subseteq \mathbf{R}_k$. This shows that $\mathbf{R}_* \subseteq \cup \mathbf{R}_k$, $k \ge 0$, and our claim is proved. Now fix $k \ge 0$. Since $((I^n)^{k+1} : (I^n)^k) = (I^n(I^n)^k : (I^n)^k)$, and since $(I^n)^k \in \Delta$, we have $((I^n)^{k+1} : (I^n)^k) \subseteq (I^n)_\Delta$. Thus, $\mathbf{R}_k \subseteq \mathbf{R}_\Delta$. It follows that $\mathbf{R}_* \subseteq \mathbf{R}_\Delta$, so that $\mathbf{R}_*$ is a finite $\mathbf{R}$-module. Since $\mathbf{R}_*$ is the union of the ascending chain of $\mathbf{R}$-submodules $\mathbf{R}_k$, $k \ge 0$, for a sufficiently large k we must have $\mathbf{R}_* = \mathbf{R}_k$. It is clear that (e) holds for this k.

(13.16) Remark. We give a partial converse to (13.15)(b). Suppose that $K \in \Delta$ is regular, and $(I^n)_\Delta = (I^n K : K)$ for all $\mathbf{n} \in \mathbf{N}_g$. We claim that $\mathbf{R}_\Delta$ is a finite $\mathbf{R}$-module. We see that $K(I^n)_\Delta \subseteq I^n K \subseteq I^n$. Therefore, $K\mathbf{R}_\Delta \subseteq \mathbf{R}$. If x is a regular element in K (so x is regular in $\mathbf{R}$), then $\mathbf{R}_\Delta \subseteq \mathbf{R}x^{-1}$, proving our claim.

14 GRADE FUNCTIONS

Motivated by the existence of classical, asymptotic and essential grade, in this chapter we look at an abstract concept of grade function.

Notation. S will be an arbitrary multiplicatively closed subset of R (with $0 \notin S$), and I_S will be an arbitrary proper ideal of R_S. $A(I_S)$ will represent a subset of Spec R_S. Here, we consider $A = A(I_S)$ to be a function whose domain is the set of all proper ideals I_S over all localizations R_S of R.

Definition. A will be called a proto-grade scheme on R if for all I_S, (i) $A(I_S)$ is a finite nonempty subset of Spec R_S, (ii) $P_S \in A(I_S)$ implies $I_S \subseteq P_S$, and (iii) if $P \in$ Spec R with $P \cap S = \varnothing$, then $P \in A(I)$ if and only if $P_S \in A(I_S)$.

Definition. If A is a proto-grade scheme for R, then the sequence of elements $x_1, ..., x_n$ in R_S is called an avoiding sequence for A if $(x_1, ..., x_n)R_S \neq R_S$ and for $i = 1, ..., n$, $x_i \notin \cup\{P_S \in A((x_1, ..., x_{i-1})R_S)\}$. (For $i - 1 = 0$, we adopt the convention that the void sequence is an avoiding sequence for any proto-grade scheme, and it generates the zero ideal.) Furthermore, if $x_1, ..., x_n$ is an avoiding sequence for A with each x_i in I_S, and if $I_S \subseteq \cup\{P_S \in A((x_1, ..., x_n)R_S)\}$, then $x_1, ..., x_n$ will be called a maximal avoiding sequence for A in I_S.

It follows from the prime avoidance lemma and part (ii) of the next lemma that any avoiding sequence for a proto-grade scheme A in I_S can be extended to a maximal avoiding sequence for A in I_S.

(14.1) Lemma. Let A be a proto-grade scheme on R.

i) If P_S is a prime minimal over I_S, then $P_S \in A(I_S)$.

ii) If $x_1, ..., x_n$ is an avoiding sequence for A in R_S, then
height $(x_1, ..., x_n)R_S = n$.

Proof. For (i), as I_P is a proper ideal in R_P, the definition shows that
$A(I_P)$ is nonempty, and every prime in $A(I_P)$ contains I_P. As P_P is the
only prime in R_P containing I_P, we see that $A(I_P) = \{P_P\}$. Now by
definition, $P_P \in A(I_P)$ implies $P \in A(I)$, which in turn implies
$P_S \in A(I_S)$. This proves (i). (ii) follows from (i) and the principal ideal
theorem.

Definition. If A is a proto-grade scheme on R, and if for all I_S, any two
maximal avoiding sequences for A in I_S have the same length (say n),
then A will be called a grade scheme on R, and an avoiding sequence
for A will be called an A-sequence. Furthermore, the length n of any
maximal A-sequence in I_S will be called the A-grade of I_S. The
function $f(I_S) = n$ will be called the grade function of A.

Definition. A function f from the set of all I_S will be called a grade
function on R if it is the grade function of some grade scheme A on R.
Such an A will be called a grade scheme for f.

Let us give some examples. If $A(I_S) = E(I_S)$, the set of
essential primes of I_S, then A is a grade scheme for R, and it is easily
seen that an A-sequence is exactly an essential sequence. Thus, the
grade function for $E(I_S)$ is the essential grade function. On the other
hand, if we let $A(I_S) = Q(I_S)$, the quintessential primes of I_S, we get (in
general) a different grade scheme on R, and in this case, the
A-sequences are exactly the quinessential sequences mentioned in
(3.15). However, as (3.15) points out, quintessential sequences are
identical to essential sequences, and so the grade function for $Q(I_S)$ is
again the essential grade function. This illustrates that two different
grade schemes on R may give rise to the same grade function.
Similarly, (3.15) shows that the grade schemes $\overline{Q}^*(I_S)$ and $\overline{A}^*(I_S)$
both have asymptotic grade as their grade function, and the two grade

schemes Ass R_S/I_S and $A^*(I_S)$ both have classical grade as their grade function. As a final example, let $A(I_S)$ be the set of primes minimal over I_S. The grade function of this grade scheme is the height function. We also mention that in general, $A(I) = $ Ass $R/\overline{I}$ is a proto-grade scheme (easily seen) but is not a grade scheme. The example in [M2, pp. 40-41] gives a local domain (R, M) and a nonzero element a of R, such that $M \in \overline{A}^*((a)) - $ Ass $R/\overline{(a)}$. Thus some power of a is a maximal avoiding sequence for A in M, while a itself is a nonmaximal avoiding sequence for A in M.

The definition of a grade scheme is rather cumbersome, and hence, so is the definition of a grade function. The main result of this chapter, to which we now turn, is that grade functions can be characterized in a useful manner which does not mention grade schemes. The following remark, which shows that we can usually avoid having to talk about localizations of R, will be tacitly used in many of our discussions.

(14.2) Remark: Let A be a grade scheme (or proto-grade scheme) on R, and let T be a localization of R. If I_S is an ideal in some localization T_S of T, then since T_S is also a localization of R, $A(I_S)$ exists, and we easily see that A restricted to {I_S | I_S is an ideal in some localization T_S of T} is a grade scheme (or proto-grade scheme) on T. Therefore, we will routinely state results for R, knowing that they automatically hold for any localization of R as well. Furthermore, in proofs we will often simplify notation by acting as if we are working within R itself, rather then some localization of R. This will be particularly useful when dealing with some avoiding sequence for A, which in general may come from some localization R_S. We will affect this simplification by making the following notational convention, which surpresses reference to what localization is under consideration.

Notation. If A is a proto-grade scheme on R, and if $x_1, ..., x_n$ is a sequence of elements in some localization R_S of R, with $(x_1, ..., x_n)R_S \neq R_S$, then we will use $A(x_1, ..., x_n)$ to denote $A((x_1, ..., x_n)R_S)$.

(14.3) Lemma. Let A be a grade scheme for the grade function f.

i) Let $x_1, ..., x_n$ be an A-sequence in R, and let $(x_1, ..., x_n)R \subseteq$
$P \in$ Spec R. Then $P \in A(x_1, ..., x_n)$ if and only if $f(P_P) = n$.

ii) For any ideal I of R, $f(I) = \min\{f(P_P) \mid I \subseteq P \in$ Spec R$\}$.

iii) For any ideal I of R, $f(I) \leq$ height I.

Proof. i) Since by definition, proto-grade schemes behave well with
respect to localization, and since $(x_1, ..., x_n)R_P \neq R_P$, it is easily seen
that $x_1, ..., x_n$ is an A-sequence in P_P (i.e., its image in R_P is an
A-sequence in P_P). Now $f(P_P) = n$ if and only if $x_1, ..., x_n$ is a
maximal A-sequence in P_P if and only if $P_P \in A((x_1, ..., x_n)R_P)$ if
and only if $P \in A(x_1, ..., x_n)$.

ii) If $I \subseteq P \in$ Spec R, then obviously $f(I) \leq f(P)$, since any A-sequence in I
is also an A-sequence in P. furthermore, any A-sequence in P is also
an A-sequence in P_P, and so $f(P) \leq f(P_P)$. Thus $f(I) \leq f(P_P)$. It only
remains to show that for some such P, we get equality. Let $x_1, ..., x_n$ be a
maximal A-sequence in I. Thus $I \subseteq \cup\{P \in A(x_1, ..., x_n)\}$, and so for
some $P \in A(x_1, ..., x_n)$, we have $I \subseteq P$. By (i), $f(P_P) = n = f(I)$.

iii) If $f(I) = n$, then I contains an A-sequence of length n, which by
(14.1)(ii), generates an ideal of height n, so that height $I \geq n$.

                    We come to our main result.

(14.4) Theorem. Let f be a nonnegative integer valued function defined
on the set $\{I_S \mid I_S$ is an ideal in some localization R_S of R$\}$. Then f is a
grade function on R if and only if the following three conditions hold.

i) $f(I_S) = \min\{f(P_P) \mid I_S \subseteq P_S \in$ Spec $R_S\}$ for all I_S.

ii) $f(P_P) \leq$ height P for all $P \in$ Spec R.

iii) If (Q, U) is a conforming pair in R (as defined in chapter 3),
and if $f(P_P) \leq n$ for all $P \in U$, then $f(Q_Q) \leq n - 1$.

Proof. If f is a grade function on R, then (14.3)(ii) and (iii) (together with the comments in (14.2)) show that conditions (i) and (ii) hold. The proof that condition (iii) holds is identical to the proof of (3.8), replacing essential grade with our grade function f, and replacing $E(I_S)$ with $A(I_S)$ where A is any grade scheme for f.

Conversely, suppose that f satisfies (i), (ii), and (iii). Define $A_f(I_S) = \{P_S \in \text{Spec } R_S \mid I_S \subseteq P_S \text{ and } f(P_P) = f(I_P)\}$. We will show that A_f is a grade scheme on R whose grade function is f. First note that (i) implies that if $I_S \subseteq J_S$ are ideals in R_S, then $f(I) \le f(I_S) \le f(J_S)$. We will use inequalities of this sort repeatedly.

We will now show that A_f is a proto-grade scheme on R. We must show that each $A_f(I_S)$ is a nonempty, finite set. As discussed in (14.2), to simplify notation we will replace R_S by R, and I_S by I. By (i), find a $P \in \text{Spec } R$ with $I \subseteq P$ and $f(I) = f(P_P)$. Then $f(I) \le f(I_P) \le f(P_P) = f(I)$. Thus, $f(P_P) = f(I_P)$, so that $P \in A_f(I)$, which therefore is nonempty. Now suppose that $A_f(I)$ is infinite. Then by (7.13), there is a conforming pair (Q, U) with $I \subseteq Q$ and $U \subseteq A_f(I)$. For all $P \in U$, since $P \in A_f(I)$ we have $f(P_P) = f(I_P)$. By condition (i), $f(I_P) \le f(Q_Q)$, so that $f(P_P) \le f(Q_Q)$ for all $P \in U$. By condition (iii), $f(Q_Q) \le f(Q_Q) - 1$. Careful examination of this last inequality reveals a contradiction, showing that $A_f(I)$ is finite. The rest of the proof that A_f is a proto-grade scheme is straightforward.

Next consider the following statement.

(a) If I is an ideal in R with $f(I) = n$, then any maximal avoiding sequence for A_f in I has length n.

We will prove that (a) is true. Notice that (a) can be applied locally (using arguments similar to those in (14.2)), so that we will have for any ideal I_S in any localization R_S of R, the length of any maximal avoiding sequence for A_f in I_S equals $f(I_S)$. This is exactly what is needed to have A_f a grade scheme whose grade functon is f. Therefore, we can complete this proof by proving (a). We will prove (a), by using induction to simultaneously prove both (a) and the following statement.

(b) If $x_1, ..., x_n$ is an avoiding sequence for A_f in R, and if $P \in A_f(x_1, ..., x_n)$, then $f(P_P) = n$.

Let $n = 0$. Suppose that $f(I) = 0$. By condition (i), there is an $I \subseteq Q \in$ Spec R with $f(Q_Q) = 0$. Since $f(0R_Q) \le f(Q_Q) = 0$, we see that $f(Q_Q) = 0 = f(0R_Q)$. Thus, by definition, $Q \in A_f(0R)$. It follows that only the void sequence is an avoiding sequence for A_f from I. This shows that (a) holds when $n = 0$. As for (b), let $P \in A_f(0R)$. We have that $f(P_P) = f(0R_P)$. However, if q_P is a minimal prime in R_P, conditions (i) and (ii) show $f(P_P) = f(0R_P) \le f(q_q) \le$ height $q = 0$. Thus $f(P_P) = 0$, and (b) holds for $n = 0$.

Now let $n > 0$, and suppose that (a) and (b) hold for any integer $0 \le m < n$. We will show that they hold for n. For (b), let $x_1, ..., x_n$ be an avoiding sequence for A_f in R, and let $P \in A_f(x_1, ..., x_n)$. If $f(P_P) < n$, then $f(P) \le f(P_P) < n$, and (a) is inductively violated. Thus $f(P_P) \ge n$. As $P \in A_f(x_1, ..., x_n)$, $f(P_P) = f((x_1, ..., x_n)R_P)$. Let q_P be a prime minimal over $(x_1, ..., x_n)R_P$. Then $f(P_P) = f((x_1, ..., x_n)R_P) \le f(q_P) \le f(q_q) \le$ height $q \le n$, using conditions (i) and (ii), and the principal ideal theorem Thus $f(P_P) = n$, and so (b) holds for n.

As for (a), let $f(I) = n$, and let $y_1, ..., y_k$ be a maximal avoiding sequence for A_f in I. We must show that $k = n$. Suppose that $k < n$. Then for any $P \in A_f(y_1, ..., y_k)$, an inductive use of (b) shows that $f(P_P) = k < n = f(I)$. Condition (i) now shows that I is not contained in P. As this holds for each of the finitely many P in $A_f(y_1, ..., y_k)$, we have that $I \not\subseteq \cup\{P \in A_f(y_1, ..., y_k)\}$, contradicting that $y_1, ..., y_k$ is a maximal avoiding sequence for A_f in I. Thus, $k \ge n$. Now condition (i) allows us to pick a prime Q containing I with $f(Q_Q) = f(I) = n$. Since $k \ge n$, we may consider $y_1, ..., y_n$, which is certainly an avoiding sequence for A_f. To prevent (a) from being inductively violated, we must have $f((y_1, ..., y_n)R) \ge n$. Therefore, $n \le f((y_1, ..., y_n)R_Q) \le f(Q_Q) = n$. Thus $f((y_1, ..., y_n)R_Q) = f(Q_Q)$, which shows that $Q \in A_f(y_1, ..., y_n)$. Since $I \subseteq Q$, we see that $y_1, ..., y_n$ is in fact a maximal avoiding sequence for A_f in I, so that $k = n$, as desired. This proves that (a) holds for n, and completes the proof of the theorem.

The grade scheme A_f is called the canonical grade scheme for f, and is studied in [M3].

(14.5) Remarks. a) (14.4)(i) shows that a grade function is determined by its values at P_P for all $P \in$ Spec R. Thus, we can define a grade function f by defining $f(P_P)$, $P \in$ Spec R, in such a way that (14.4)(ii) and (14.4)(iii) are satisfied, and then extending the definition via (14.4)(i).

b) The little height function, assigning to each P_P the length of the shortest saturated chain of primes from P_P down to a minimal prime is not a grade function. Using [N3, example 2, pp. 202-203] one can produce an example in which this function does not satisfy (14.4)(iii).

(14.6) Corollary. Let g be a grade function on R. Let $q \in$ Spec R with $g(q_q) \geq 1$. Define f by letting $f(P_P) = g(P_P)$ for any $P \in$ Spec R with $P \neq q$, while $f(q_q) = g(q_q) - 1$. Then f defines a grade function on R (via (14.5)(a)).

Proof. We must show that f satisfies conditions (ii) and (iii) of (14.4). Clearly f satisfies (ii), since g does, and f does not exceed g. As for (iii), let (Q, U) be a conforming pair in R, and suppose that $f(P_P) \leq n$ for all $P \in$ U. We need $f(Q_Q) \leq n - 1$. If $q \notin U \cup \{Q\}$, there is no problem, since g satisfies (iii), and f and g have the same value everywhere under consideration. Now suppose that $q \in$ U. Let $U' = U - \{q\}$. Clearly (Q, U') is a conforming pair. Since $q \notin U' \cup \{Q\}$, the previous case applies. Finally, if $q = Q$, then since $g(P_P) = f(P_P) \leq n$ for all $P \in$ U, we see that $g(Q_Q) \leq n - 1$. As $f(Q_Q) = g(Q_Q) - 1$, we are done.

The upshot of (14.6) is that unless Spec R is extremely simple, there will be a plethora of grade functions on R. For example, if $R = K[X, Y, Z]$, then we can define $f(P_P)$ to equal height P for all $P \in$ Spec R, except when $P = (X, Y, Z)$ we define $f(P_P) = 0$, and when $P = (X, Y)$ we define $f(P_P) = 1$. Starting with the height function, and then iterating (14.6) four times shows that f defines a grade function on R. (However, (14.6) cannot be indiscriminately iterated infinitely many times. With R as above, let $U = \{P \in$ Spec R | P is maximal and $(X, Y) \subset P\}$. If we define $f(P_P) =$ height P unless $P \in$ U, in which case

$f(P_P) = 2$, we do not get a grade function, since (14.4)(iii) fails to hold for the conforming pair ((X, Y), U).)

As just noted, on any Noetherian ring, there are usually many grade functions. However, most of them are just defined arbitrarily, as in the example in K[X, Y, Z] just mentioned. On the other hand, classical grade, asymptotic grade, essential grade, and height are special, in that they exist in all Noetherian rings. We refer to these as natural grade functions. It is interesting to speculate on what other natural grade functions exist. Let us offer a few thoughts in that direction.

(14.7). Convention. To escape having to refer to localizations of R, we make the following convention. If we wish to define a function $f(I_S)$ on the set of all ideals of R_S for all localizations of R, and if the definition for $f(I_S)$ is merely a "localization" of the definition of $f(I)$, then we shall only define $f(I)$, and let it be understood that the definition is meant to extend to all $f(I_S)$. As an illustration of this convention, if we want to define $f(I_S)$ = height I_S, we will merely say $f(I)$ = height I.

(14.8) Proposition. Let $q_1, ..., q_n$ be a finite list, possibly with repetitions, of prime ideals of R, such that every minimal prime of R appears in this list. For $1 \leq i \leq n$, let f_i be a grade function on R/q_i. For an ideal I in R, let $f(I) = \min\{f_i(P/q_i) \mid I \subseteq P \in \text{Spec } R, \text{ and } q_i \subseteq P\}$. Then f is a grade function on R (using (14.7)).

Proof. We define a function g on the set $\{P_P \mid P \in \text{Spec } R\}$ by letting $g(P_P) = \min\{f_i(P_P/(q_i)_P) \mid q_i \subseteq P\}$. (Here, note that since every minimal prime appears among the q_i, for any $P \in \text{Spec } R$ there is a $q_i \subseteq P$. Also, $P_P/(q_i)_P$ is the maximal ideal of R/q_i localized at P/q_i, and since f_i is a grade function on R/q_i, $f_i(P_P/(q_i)_P)$ exists. Thus the definition of $g(P_P)$ makes sense.)

We now show that g satisfies conditions (ii) and (iii) of (14.4). Since f_i satisfies condition (ii), we see that if $q_i \subseteq P$, then $f_i(P_P/(q_i)_P) \leq \text{height } P/q_i \leq \text{height } P$. Thus $g(P_P) \leq \text{height } P$, so that (ii) is satisfied. For condition (iii), let (Q, U) be a conforming pair in R, and suppose that for all $P \in U$, $g(P_P) \leq n$. We need $g(Q_Q) \leq n - 1$. For each $P \in U$, by definition of g, we see that there is an i with $q_i \subseteq P$ and

$f_i(P_P/(q_i)_P) \leq n$. However, U is infinite, and $1 \leq i \leq n$. Thus, there must be some one fixed i such that for infinitely many $P \in U$, we have for that i, $q_i \subseteq P$ and $f_i(P_P/(q_i)_P) \leq n$. Let U' be the subset of U consisting of those infinitely many P. It is easily seen that (Q, U') is a conforming pair. In particular, the intersection of all $P \in U'$ is Q. Thus, $q_i \subseteq Q$.

It is now easily seen tha $(Q/q_i, \{P/q_i \mid P \in U'\})$ is a conforming pair in R/q_i. Since f_i satisfies condition (iii) of (14.4), and since

$f_i(P_P/(q_i)_P) \leq n$ for all $P \in U'$, we see that $f_i(Q_Q/(q_i)_Q) \leq n - 1$.
By definition of g, $g(Q_Q) \leq n - 1$, as desired.

  For I an ideal of R, we now define $g(I) =$

$\min\{g(P_P) \mid I \subseteq P \in \text{Spec } R\}$. Then by (14.5)(a) (and (14.7)), $g(I)$ is a grade function. Since $g(P_P) = \min\{f_i(P_P/(q_i)_P) \mid q_i \subseteq P\}$, we see that $g(I) = \min\{f_i(P_P/(q_i)_P) \mid I \subseteq P \in \text{Spec } R, \text{ and } q_i \subseteq P\}$.

  We now claim that $g(I) = f(I)$, as defined in the statement of the proposition. (That is, we show that $\min\{f_i(P_P/(q_i)_P) \mid I \subseteq P \in$

Spec R and $q_i \subseteq P\} = \min\{f_i(P/q_i) \mid I \subseteq P \in \text{Spec } R \text{ and } q_i \subseteq P\}$.) Since each f_i is a grade function, we know that $f_i(P_P/(q_i)_P) \geq f_i(P/q_i)$. Therefore, $g(I) \geq f(I)$. Suppose now that the minimum defining $f(I)$ occurs at $f_i(P/q_i)$ (so $I \subseteq P$). Since f_i is a grade function on R/q_i,

(14.4)(i) shows there is a prime P'/q_i containing P/q_i such that

$f(I) = f_i(P/q_i) = f_i(P'_{P'}/(q_i)_{P'}) \geq g(I)$, since $I \subseteq P'$. Thus $g(I) = f(I)$, as claimed. Since $g(I)$ is a grade function, the proposition is proved.

  Counting the various options, the next example presents eight different natural grade functions. (The reader may wish to verify that asymptotic and essential grade are among these eight, but classical grade and the height function are not.)

(14.9) Examples. Let $q_1, ..., q_n$ be the minimal primes of R, (alternately, the set of prime divisors of zero in R). For each i, let f_i be the height function on R/q_i. (Alternately, let f_i be the asymptotic grade function, or the essential grade function, or the classical grade function on R/q_i). For I an ideal in R, let $f(I) = \min\{f_i(P/q_i) \mid I \subseteq P \in \text{Spec } R$ and $q_i \subseteq P\}$. Then (14.8) (and (14.7)) show that f is a grade function on R.

(14.10) Question. Let g be a natural grade function. For I an ideal in R, define $f(I) = \min\{g(Pp^*/q) \mid I \subseteq P \subseteq \text{Spec } R$ and q is a minimal prime (alternately, a prime divisor of zero) in Rp^*, the completion of $Rp\}$. Is f a grade function? We are particularly curious about this when g is classical grade. (If g is the height function or asymptotic grade, and we take q minimal in Rp^*, then it is not hard to verify that f is asymptotic grade. Similarly, if g is the height function or essential grade, and we take q a prime divisor of zero in Rp^*, then f is essential grade. Thus the answer to our question is yes in these cases.)

(14.11) Proposition. Let $R \subseteq T$ be Noetherian rings such that for every $P \in \text{Spec } R$, there is a $Q \in \text{Spec } T$ with $Q \cap R = P$, (i.e., lying over holds). Let g be a grade function on T. For I_S an ideal in R_S, let $f(I_S) = g(IT_S)$. Then f is a grade function on R.

Proof. First note that if S is a multiplicatively closed subset of R which is disjoint from I, then by lying over, S is also disjoint from IT, so that $g(IT_S)$ is defined. Since g satisfies (14.4)(i), for $P \in \text{Spec } R$ we easily see that $f(Pp) = \min\{g(Q_Q) \mid Q \in \text{Spec } T$ and $Q \cap R = P\}$. From this, we see that f satisfies condition (i) of (14.4). We claim it also satisfies conditions (ii) and (iii) of that result. For (ii), let $P \in \text{Spec } R$ and let $Q \in \text{Spec } T$ with $Q \cap R = P$. Then $f(Pp) \leq g(Q_Q) \leq \text{height } Q$, since g satisfies condition (ii). We can show that f satisfies (ii) by showing that we may pick Q such that height $Q \leq$ height P. To do this, let $S = R - P$, and let $a_1, ..., a_n$ be a system of parameters in Rp (so that height $P = n$). By lying over, $(a_1, ..., a_n)T_S \neq T_S$. Let $Q_S \in \text{Spec } T_S$ with Q_S minimal over $(a_1, ..., a_n)T_S$. Clearly $Q \cap R = P$, and by the principal ideal theorem, we have height $Q \leq n = $ height P. Thus, f satisfies condition (ii) of (14.4).

To show that f satisfies condition (iii) of (14.4), let (q, U) be a conforming pair in R, and assume that $f(P_p) \leq n$ for all $P \in U$. It is easy to see that we may pick a set W of primes in T such that for each $P \in U$, there is a unique $Q \in W$ with $Q \cap R = P$, and $g(Q_Q) = f(P_p) \leq n$. By (7.13), there is a conforming pair (q', U') in T such that $qT \subseteq q'$, and $U' \subseteq W$. Since g satisfies (14.4)(iii), clearly $g(q'_{q'}) \leq n - 1$. Now the intersection of all the primes in U' is q'. Also, $\{Q \cap R \mid Q \in U'\}$ is an infinite subset of U, and so the intersection of this set of primes is q. Therefore, $q' \cap R = q$. As $g(q'_{q'}) \leq n - 1$, we have $f(q_q) \leq n - 1$. This shows that f satisfies (14.4)(iii). According to (14.4), f is a grade function on R.

(14.12) Remark. Let R, T, f and g be as in (14.11). The reader should have little trouble verifying the following facts. (a) If T is a faithfully flat extension of R, and if g is any of classical, essential, or asymptotic grade, then so is f. (b) If T is an integral extension of R such that every minimal prime of T contracts to a minimal prime of R, and if g is asymptotic grade, then so is f. (c) If T is a finite module extension of R such that every prime in Ass T contracts to a prime in Ass R, and if g is essential grade, then so is f.

(14.13) Example. Let R be a semi-local ring with completion R^*. For I an ideal of R, let f(I) = min{classical grade P/q | $IR^* \subseteq P \in$ Spec R^*, and q is a minimal prime (alternately, a prime divisor of zero) in R^* with $q \subseteq P$}. Then f is a grade function on R. To see this, apply (14.8) to R^*, and then apply (14.11) to $R \subseteq R^*$.

(14.14) Question. Can the grade function defined in (14.13) be extended to a natural grade function existing on all Noetherian rings. If so, is the extention given by the function defined in (14.10) (using classical grade for g)?

(14.15) Remark. Let T be a not necessarily Noetherian ring with $R \subseteq T \subseteq R'$. For I an ideal in R, define f(I) = height IT. We leave it to the reader to verify that f is a grade function on R. (To show that

(14.4)(iii) holds, it is useful to know that if $Q \in$ Spec R, then only finitely many primes of T are minimal over QT. This follows from [M2. Corollary 1.8] extended to nondomains via (1.1)(c).) We wonder if the case $T = R^e$ (as in Ch. 7) is of particular interest? ($R \subseteq R^e \subseteq R'$ by [M2, Proposition 10.21].)

        The examples at the start of this chapter show that two different grade schemes can have the same grade function. We consider this a bit further, with the aim to showing that any grade function has a "minimum" grade scheme.

(14.16) Lemma. Let A be a grade scheme on R. Let B be a proto-grade scheme on R. The following are equivalent.

a) B is a grade scheme on R with the same grade function as A.

b) $A(x_1, ..., x_n) = B(x_1, ..., x_n)$ for all A-sequences $x_1, ..., x_n$.

c) A-sequences are identical to avoiding sequences for B.

Proof. a) $\Rightarrow$ c) Suppose B is a grade scheme having the same grade function (say f) as A. As B is a grade scheme, an avoiding sequence for B is a B-sequence. We must show that A-sequences and B-sequences are identical. We will induct on the length of the sequence in question. Suppose the result is true for any sequence of length n - 1, the case n - 1 = 0 being true by convention. Let us consider a sequence $x_1, ..., x_n$ which is (let us say) an A-sequence. We must show it is also a B-sequence. Since $x_1, ..., x_{n-1}$ is an A-sequence, by induction it is also a B-sequence. Let $P \in B(x_1, ..., x_{n-1})$. We must only show that $x_n \notin P$. However, (14.3)(i) applied to B shows that $f(P_P) = n - 1$. That same result applied to A now shows that $P \in A(x_1, ..., x_{n-1})$. As $x_1, ..., x_n$ is an A-sequence, we must have $x_n \notin P$, as desired.

c) $\Rightarrow$ a). This is straightforward from the definitions.

a) $\Rightarrow$ b) Let (a) (and hence (c)) hold, and let $x_1, ..., x_n$ be an A-sequence. By (c), it is also a B-sequence. Now (14.3)(i) shows that $A(x_1, ..., x_n)$

consists exactly of those primes P which contain $x_1, ..., x_n$ and have $f(P_P) = n$, where f is the common grade function of A and B. However, this also describes $B(x_1, ..., x_n)$, and so $A(x_1, ..., x_n) = B(x_1, ..., x_n)$.

b) $\Rightarrow$ c) Suppose (b) holds. We will show that a sequence $y_1, ..., y_m$ is an avoiding sequence for B if and only if it is an A-sequence. We induct on m, the case m = 0 being trivial. Assume that our claim is true for sequences of length m - 1. Let $y_1, ..., y_m$ be either an avoiding sequence for B, or an A-sequence. Then the same can be said of $y_1, ..., y_{m-1}$, and so by induction we see that $y_1, ..., y_{m-1}$ is both an avoiding sequence for B and an A-sequence. In particular, by (b), $A(y_1, ..., y_{m-1}) = B(y_1, ..., y_{m-1})$. Since $y_1, ..., y_m$ is either an avoiding sequence for B or an A-sequence, no prime in $A(y_1, ..., y_{m-1}) = B(y_1, ..., y_{m-1})$ can contain y_m. Therefore we see that $y_1, ..., y_m$ is both an avoiding sequence for B and an A-sequence. This proves (c).

Note that (14.16) shows that if A and B are two grade schemes on R having the same grade function f, then A-sequences are identical to B-sequences. Thus, the next definition is unambiguous.

Definition. Let f be a grade function on R. The sequence $x_1, ..., x_n$ is an f-sequence, if it is an A-sequence for any grade scheme A having f as its grade function.

Notation. v(I) will be the smallest number of elements in a generating set for I.

(14.17) Lemma. Let f be a grade function on R. Let I be an ideal of R with f(I) = v(I). Then I can be generated by an f-sequence, necessarily of length f(I).

Proof. The proof that I can be generated by an f-sequence is similar to the proof of [Kp, Theorem 125]. The length of that f-sequence is at least as great as v(I) = f(I), but of course f(I) is as least as great as the length of that f-sequence. Thus equality holds.

Notation. Let A and B be two proto-grade schemes on R. Say $A \subseteq B$ if $A(I_S) \subseteq B(I_S)$ for all ideals I_S in all localizations R_S of R.

We now show that any grade function on R has a minimum grade scheme under the partial ordering given by $\subseteq$.

(14.18) Proposition. Let f be a grade function on R. For I_S an ideal in R_S, define $A_f{}^m(I_S) = \{P_S \in \text{Spec } R_S \mid I_S \subseteq P_S$ and either P_S is minimal over I_S or $f(P_P) = f(I_P) = v(I_P)\}$. Then $A_f{}^m$ is a grade scheme for f, and if A is any grade scheme for f, then $A_f{}^m \subseteq A$.

Proof. Let A be any grade scheme for f. Suppose that $P_S \in A_f{}^m(I_S)$. If P_S is minimal over I_S, then by (14.1)(i), $P_S \in A(I_S)$. If P_S is not minimal over I_S, then by definition of $A_f{}^m$, we must have that $f(P_P) = f(I_P) = v(I_P)$. By (14.17), I_P can be generated by an f-sequence $x_1, ..., x_n$, with $n = f(I_P) = f(P_P)$. Now $x_1, ..., x_n$ is an A-sequence in P_P, and it must be a maximal A-sequence in P_P, since $f(P_P) = n$. Therefore, we must have $P_P \in A((x_1, ..., x_n)R_P)$. That is, $P_P \in A(I_P)$. It follows from the definition of proto-grade scheme, that $P_S \in A(I_S)$. Therefore, we have shown that for all I_S, $A_f{}^m(I_S) \subseteq A(I_S)$.

Since primes minimal over I_S are in $A_f{}^m(I_S)$, and since $A_f{}^m(I_S) \subseteq A(I_S)$, we see that $A_f{}^m(I_S)$ is a nonempty finite set. We now can easily verify that $A_f{}^m$ is a proto-grade scheme. It remains to show that $A_f{}^m$ is a grade scheme having grade function f. To do this, we use (14.16)(b)$\Rightarrow$(a). Let $x_1, ..., x_n$ be an A-sequence. We want that $A(x_1, ..., x_n) = A_f{}^m(x_1, ..., x_n)$. The first paragraph of this proof shows that $A_f{}^m(x_1, ..., x_n) \subseteq A(x_1, ..., x_n)$. Therefore, let $P \in A(x_1, ..., x_n)$. As $x_1, ..., x_n$ is still an A-sequence in R_P, clearly $n \leq f((x_1, ..., x_n)R_P) \leq f(P_P) = n$, the equality by (14.3)(i). Also, using the principal ideal theorem and (14.3)(iii), we see that $n \geq v((x_1, ..., x_n)R_P) \geq$ height $(x_1, ..., x_n)R_P \geq f((x_1, ..., x_n)R_P) \geq n$. Together, these show that $n = f(P_P) = f((x_1, ..., x_n)R_P) = v((x_1, ..., x_n)R_P)$. By definition, $P \in A_f{}^m(x_1, ..., x_n)$.

The material in this chapter taken is from [M3].

15 PARTIAL ORDERINGS ON GRADE FUNCTIONS

Notation. Throughout this chapter, W will be a nonempty set of grade functions on R. For $f \in$ W, let B_f be a grade scheme for f. We define $\cup B_f$, $f \in$ W, and $\cap B_f$, $f \in$ W, as follows. For I_S any ideal in any localization R_S of R, let $(\cup B_f)(I_S) = \cup\{B_f(I_S) \mid f \in W\}$ and $(\cap B_f)(I_S) = \cap\{B_f(I_S) \mid f \in W\}$. (In general, $\cup B_f$ and $\cap B_f$ will not be grade schemes.) Recall that if B and C are grade schemes on R, $B \subseteq C$ was defined in the prior to (14.18).

This chapter will focus on the following questions.

Question 1. Does W have a least upper bound in the set of grade functions on R?

Question 2. If for each $f \in$ W, B_f is a grade scheme for f, when is $\cap B_f$, $f \in$ W, a grade scheme, and if so, for what grade function?

Question 3. Does W have a greatest lower bound in the set of grade functions on R?

Question 2. If for each $f \in$ W, B_f is a grade scheme for f, when is $\cup B_f$, $f \in$ W, a grade scheme, and if so, for what grade function?

Of course the function theoretic least upper bound and greatest lower bound for W, max W and min W, always exist, but as we next show, they may not be grade functions. Thus, their existence does not answer questions 1 and 3.

(15.1) Examples. We show that max W and min W may not be grade functions. First, let R be a 1-dimensional domain with exactly two maximal ideals, M and N. Define the grade function f by specifying that $f(M_M) = 1$ and $f(N_N) = 0$ (and using (14.5)(a)). Define the grade function h by specifying $h(M_M) = 0$ and $h(N_N) = 1$. By (14.4)(i),

$f(M \cap N) = \min\{f(M_M), f(N_N)\} = 0$. Similarly, $h(M \cap N) = 0$.

Now let $k = \max\{f, h\}$. Then $k(M \cap N) = 0$. Also, $k(M_M) = \max\{f(M_M),$ $h(M_M)\} = 1$, and similarly, $k(N_N) = 1$. Thus $k(M \cap N) \neq \min\{k(M_M),$ $k(N_N)\}$, and so k does not satisfy (14.4)(i). This shows that $k = \max\{f, h\}$ is not a grade function.

Next, let R be a 2-dimensional domain, let Q be a height 1 prime of R, and let P_1, P_2, P_3, ... be infinitely many height 2 primes each of which contains Q. For i = 1, 2, 3, ... define the grade function f_i by specifying that $f_i(q_q)$ = height q for all q ∈ Spec R - $\{P_i\}$, while for q = P_i, let $f_i(q_q) = 1$. By (14.6), each f_i does determine a grade function. Let $g = \min\{f_i \mid i = 1, 2, 3, ...\}$. If $U = \{P_1, P_2, P_3, ...\}$, then

(Q, U) is a conforming pair in R. Clearly $g(P_P) = 1$ for all P ∈ U. However, $g(Q_Q) = 1$ as well. Thus g fails to satisfy (14.4)(iii), and so is not a grade function. (Incidentally, W does have a grade function greatest lower bound. It is the grade function h determined by letting $h(q_q)$ = height q for all q ∈ Spec R - ($\{Q\} \cup U$), while $h(q_q) = 1$ for all q ∈ U, and finally, $h(Q_Q) = 0$.)

We now show that question 1 is easily answered.

(15.2) Proposition. For all P ∈ Spec R, define $g(P_P) =$

$\max\{f(P_P) \mid f \in W\}$, and extend g via (14.5)(a). Then g is a grade function on R and g is the least upper bound of W.

Proof. If g is a grade function, then (14.4)(i) easily shows that it is the least upper bound of W. To show that g is a grade function, we must show that it satisfies conditions (ii) and (iii) of (14.4). It is trivial that (ii) holds for g, since it holds for each f ∈ W. For condition (iii), let (Q, U) be a conforming pair in R, and suppose that $g(P_P) \leq n$ for all

P ∈ U. We need $g(Q_Q) \leq n - 1$. Since $f(P_P) \leq g(P_P) \leq n$ for all f ∈ W and all P ∈ U, and since each f ∈ W satisfies (14.4)(iii), we see that $f(Q_Q) \leq n - 1$. Thus by definition, $g(Q_Q) \leq n - 1$.

Let us turn to question 2. If for each $f \in W$, B_f is a grade scheme for f, the next two examples show that (even when W is finite) $\cap B_f$, $f \in W$, need not be a grade scheme, and even if it is a grade scheme, it need not be a grade scheme for the least upper bound of W.

(15.3) Example. Let (R, M) be a 2-dimensional local domain, and let Q be a height 1 prime of R. We define grade functions f and g on R by specifying that $f(M_M) = 1$, and $f(P_P)$ = height P for all other primes in R, while $g(M_M) = 1$, $g(Q_Q) = 0$, and $g(P_P)$ = height P for all other primes in R. Starting with the height function, and then making iterated use of (14.6) shows that f and g define grade functions via (14.5)(a). By (14.4)(i), clearly $f \geq g$. Let B and C be grade schemes for f and g respectively, with $C \subseteq A_g$, the canonical grade scheme for g. We claim that $B \cap C$ is not a grade scheme. (It is a proto-grade scheme.) By (14.3)(i) (and the fact that the void sequence is a C-sequence), we see that $C(0) = \{0, Q\}$. Similarly, $B(0) = \{0\}$. Clearly $(B \cap C)(0) = \{0\}$, and so any nonzero element $x \in M$ is an avoiding sequence for $B \cap C$. If also $x \notin Q$, then x is both an f-sequence and a g-sequence. By (14.3)(i), since $f(M_M) = 1 = g(M_M)$, we have $M \in B(xR) \cap C(xR) = (B \cap C)(xR)$. Thus, any $x \in M - Q$ is a maximal avoiding sequence for $B \cap C$ in M. On the other hand, suppose that $0 \neq x \in Q$. Since $g(Q_Q) = 0$, (14.4)(i) shows $g(xR) = 0$. As $R = R_M$, $g(xR_M) = 0 \neq 1 = g(M_M)$. By definition, $M \notin A_g(xR)$, and since we are assuming that $C \subseteq A_g$, we have $M \notin C(xR)$. Thus $M \notin (B \cap C)(xR)$. This shows that x is not a maximal avoiding sequence for $B \cap C$ in M. Therefore, we can find avoiding sequences for $B \cap C$ of different lengths in M, proving the claim that $B \cap C$ is not a grade scheme.

(15.4) Example. Let (R, M) be a two dimensional local domain. Define f and h by specifying $f(P_P) = h(P_P)$ = height P for all primes $P \neq M$, while $f(M_M) = 1$ and $h(M_M) = 0$. By (14.6), f and h determine grade functions. Clearly $f \geq h$. We easily see that $A_f{}^m \cap A_h{}^m$ assigns to any ideal the primes minimal over that ideal. Thus $A_f{}^m \cap A_h{}^m$ is a grade scheme for the height function, not for f, the least upper bound of $\{f, h\}$.

In view of the preceding two examples, one does not expect to be able to say much in the way of positive results concerning question 2. After the next lemma, we do offer one positive result.

(15.5) Lemma. Let f and h be grade functions on R, with grade schemes B and C respectively. The following are equivalent.

i) $f \geq h$.

ii) Any h-sequence $x_1, ..., x_n$ is also an f-sequence.

iii) For any h-sequence $x_1, ..., x_n$, $B(x_1, ..., x_n) \subseteq C(x_1, ..., x_n)$.

Proof. (iii) $\Rightarrow$ (ii) $\Rightarrow$ (i) is easy. Assuming (i), we will simultaneously prove (ii) and (iii), by inducting on n. For n = 0, (ii) is trivial. As for (iii), if $P \in B(0R)$, then $f(P_P) = 0$, so that by (i), $h(P_P) = 0$, showing that $P \in C(0R)$. Now suppose that we know (ii) and (iii) both hold for n - 1. Together, these imply that (ii) holds for n, as is easily seen. As for (iii), let $P \in B(x_1, ..., x_n)$. By (ii), $x_1, ..., x_n$ is an f-sequence, so that $f(P_P) = n$, using (14.3)(i). By (i), $h(P_P) \leq n$. Since P_P contains the h-sequence $x_1, ..., x_n$, $h(P_P) = n$, so that $P \in C(x_1, ..., x_n)$.

(15.6) Proposition. Let W be finite, and for each $f \in W$ let B_f be a grade scheme for f with $A_f \subseteq B_f$. If $\cap B_f, f \in W$, is a grade scheme, then it is a grade scheme for the least upper bound of W.

Proof. Let k be the least upper bound of W (which we know exists by (15.2)), and suppose that $B = \cap B_f, f \in W$, is a grade scheme for the grade function g. Since $B \subseteq B_f$, clearly $g \geq f$ for all $f \in W$, so that $g \geq k$. Now let $P \in$ Spec R. We will prove the result by showing that $k(P_P) = g(P_P)$. Let $k(P_P) = n$, so that $n = \max\{f(P_P) \mid f \in W\}$. We claim we can find a sequence of elements $x_1, ..., x_n$ in P_P such that if for any $f \in W$, $f(P_P) = r$, then $x_1, ..., x_r$ is an f-sequence. To do this, by prime avoidance pick x_1 in P_P but not in any prime in $A_f(0R_P)$ for any $f \in W$ having $f(P_P) > 0$. Thus if $f \in W$ and $f(P_P) > 0$, then x_1 is an f-sequence. Now pick x_2 in P_P but not in any prime in $A_f(x_1 R_P)$ for

any $f \in W$ having $f(P_P) > 1$. Thus x_1, x_2 is an f-sequence for any $f \in W$

with $f(P_P) > 1$. Iterating this process proves the claim. For $f \in W$, if

$f(P_P) = r$, then since x_1, ..., x_r is an f-sequence we have

$f(P_P) = r \leq f(x_1, ..., x_r) \leq f(x_1, ..., x_n) \leq f(P_P)$. Equality holds

throughout, showing that $P_P \in A_f(x_1, ..., x_n)$. However, by assumption,

$A_f \subseteq B_f$, so that $P_P \in B_f(x_1, ..., x_n)$. This is true for all $f \in W$,

so $P_P \in B(x_1, ..., x_n)$. Now for some $f \in W$, $n = k(P_P) = f(P_P)$, and for

that f, x_1, ..., x_n is an f-sequence. We already have $g \geq k \geq f$, so (15.5)

shows that x_1, ..., x_n is also a g-sequence. Since $P_P \in B(x_1, ..., x_n)$ and

B is a grade scheme for g, (14.3)(i) shows that $g(P_P) = n$, as desired.

          We now consider questions 3 and 4. We start with the case

that W is finite.

(15.7) Proposition. Let W be finite. Then min W is the greatest lower

bound of W. Also, if for each $f \in W$, B_f is a grade scheme for f, then

$\cup B_f$, $f \in W$, is a grade scheme for min W.

Proof. For the first part, clearly it is enough to show that min W is a

grade function. Let $h(P_P) = \min\{f(P_P) \mid f \in W\}$. We will show that h

satisfies conditions (ii) and (iii) of (14.4). Clearly h satisfies (ii), since

each $f \in W$ does. Let (Q, U) be a conforming pair in R, with $h(P_P) \leq n$

for all $P \in U$. Since U is infinite, while W is finite, the definition of

$h(P_P)$ shows that for some single fixed $f \in W$, there is an infinite subset

U' of U such that for all $P \in U'$, $f(P_P) \leq n$. As (Q, U') is a conforming

pair and f is a grade function, we have that $f(Q_Q) \leq n - 1$. Thus

$h(Q_Q) \leq f(Q_Q) \leq n - 1$. This shows that h satisfies (14.4)(iii). Using

(14.5)(a), h determines a grade function on R. Now for I_S an ideal in

some localization R_S of R, $h(I_S) = \min\{h(P_P) \mid I_S \subseteq P_S \in \operatorname{Spec} R_S\} =$

$\min\{f(P_P) \mid I_S \subseteq P_S \subseteq \operatorname{Spec} R_S, f \in W\} = \min\{f(I_S) \mid f \in W\}$.

Thus h = min W, and so min W is a grade function. This proves the

first part of the result.

Now let $B = \cup B_f$, $f \in W$, and let A_h be the canonical grade scheme of $h = \min W$. We will show that B is a grade scheme for h by using $(14.16)(a) \Leftrightarrow (b)$. It is easy to see that B is a proto-grade scheme. Let $x_1, ..., x_n$ be an A_h-sequence. Since for $f \in W$, $f \geq h$, (15.5) shows that $B_f(x_1, ..., x_n) \subseteq A_h(x_1, ..., x_n)$. Thus $B(x_1, ..., x_n) \subseteq A_h(x_1, ..., x_n)$. For the reverse inclusion, if $P \in A_h(x_1, ..., x_n)$, then by $(14.3)(i)$, $h(P_P) = n$. By the previous paragraph, for some $f \in W$, $f(P_P) = n$. However, since $f \geq h$, (15.5) shows that $x_1, ..., x_n$ is an f-sequence. Therefore $(14.3)(i)$ gives $P \in B_f(x_1, ..., x_n) \subseteq B(x_1, ..., x_n)$. By (14.16), B is a grade scheme for $h = \min W$.

(15.8) Corollary. Let f and g be grade functions on R. Then $f \geq g$ if and only if there are grade schemes B and C for f and g respectively, with $B \subseteq C$. In fact, if $f \geq g$, and if B is any grade scheme for f, then there is a grade scheme C for g with $B \subseteq C$.

Proof. Surely if B and C are grade schemes for f and g respectively, and if $B \subseteq C$, then $f \geq g$. Conversely, suppose that $f \geq g$, and that B is any grade scheme for f. Then (15.7) shows that $B \cup A_g$ is a grade scheme for g, which we may take for our desired C.

We now look at questions 3 and 4 for arbitrary W. We first note that when W is infinite, it may not have any lower bounds at all. To see this, use the second example in (15.1) with the variation that for $q = P_i$, define $f_i(q_q) = 0$. If k were a lower bound for W, we would have to have $h(Q_Q) < 0$, which is not allowed.

(15.9) Proposition. W has a greatest lower bound if and only if it has some lower bound.

Proof. As one direction is trivial, suppose that W has a lower bound. Let W' be the set of all lower bounds to W. By (15.2), W' has a least upper bound g. From the description of $g(P_P)$ given in (15.2), it is easily seen that g is a lower bound for W, and so must be the greatest lower bound for W.

As for question 4, it may happen that $\cup B_f$, $f \in$ W, is not a grade scheme. In the second example in (15.1), we saw that min W is not a grade function. Thus (15.10) shows that for that example, $\cup B_f$, $f \in$ W, cannot be a grade scheme.

(15.10) Proposition. For each $f \in$ W, let B_f be a grade scheme for f. The following three statements are equivalent.

a) $\cup B_f$, $f \in$ W, is a grade scheme.

b) $\cup B_f$, $f \in$ W, is a grade scheme for min W.

c) For any ideal I of R, $\cup (B_f(I) \mid f \in$ W} is finite.

Furthermore, the above statements imply the next statement, and are equivalent to it in the case that for each $f \in$ W, $B_f \subseteq A_f$, the characteristic grade scheme for f.

d) For every subset W$'$ of W, min W$'$ is a grade function.

Proof. Clearly (b) $\Rightarrow$ (a) $\Rightarrow$ (c). We now show (a) $\Rightarrow$ (b). Thus, suppose that $B = \cup B_f$, $f \in$ W, is a grade scheme, and let k be its grade function. Also, let h = min W. Since $B_f \subseteq B$ for all $f \in$ W, clearly $f \geq k$. Thus $h \geq k$. Now let I be an ideal, and let $x_1, ..., x_n$ be a maximal k-sequence from I. Obviously $x_1, ..., x_n$ is also an f-sequence for all $f \in$ W. However, since our sequence is a maximal k-sequence, there is a

$P \in B(x_1, ..., x_n) = (\cup B_f)(x_1, ..., x_n)$ with $I \subseteq P$. Since for some

$f \in$ W, $P \in B_f(x_1, ..., x_n)$, we see that for that f, $x_1, ..., x_n$ is a maximal f-sequence. Therefore, $k(I) = n = f(I) \geq h(I)$. Combining this with our earlier inequality shows that k = h, proving (b). Next, we show that

(c) $\Rightarrow$ (a). If (c) holds, it is easily seen that $B = \cup B_f$, $f \in$ W, is a proto-grade scheme. Let I be an ideal, and $x_1, ..., x_n$ and $y_1, ..., y_m$ be two maximal avoiding sequences for B in I. To show that B is a grade scheme, we want n = m. By maximality, we see that there is a

$P \in B(x_1, ..., x_n)$ with $I \subseteq P$. For some $f \in$ W, we have

$P \in B_f(x_1, ..., x_n)$. Similarly, for some $g \in W$ and $Q \in B_g(y_1, ..., y_m)$, we have $I \subseteq Q$. Now clearly $x_1, ..., x_n$ is an avoiding sequence for $B_f \cup B_g$, and the existence of P shows that it is a maximal avoiding sequence for $B_f \cup B_g$ in I. Similarly, $y_1, ..., y_m$ is a maximal avoiding sequence for $B_f \cup B_g$ in I. By (15.7), we see that $B_f \cup B_g$ is a grade scheme, and so $n = m$, as desired. We now prove that (c) $\Rightarrow$ (d). If (c) holds for W, then it holds for any subset W' of W. Thus (b) holds for W', and so min W' must be a grade function, proving that (d) holds.

We now add the assumption that for each $f \in W$, $B_f \subseteq A_f$. Under this assumption, we will show that (d) $\Rightarrow$ (c). Suppose to the contrary, that (d) holds, but that (c) fails for some ideal I. That failure, together with $B_f \subseteq A_f$, shows that $\cup\{A_f(I) \mid f \in W\}$ is an infinite set of prime ideals. By (7.13), there is a conforming pair (Q, U) with $I \subseteq Q$ and $U \subseteq \cup\{A_f(I) \mid f \in W\}$. We now consider all pairs (U', W') where $U' \subseteq U$, $W' \subseteq W$, and there is a one-to-one correspondence between U' and W' such that if $P \in U'$ corresponds to $f \in W'$, then $P \in A_f(Q)$. Surely such pairs exist, since $(\varnothing, \varnothing)$ is one such. We claim that we can find such a pair with U' infinite. If not, let (U', W') be such a pair with U' as large as possible. Since U' and hence W' are finite, $\cup\{A_f(Q) \mid f \in W'\}$ is finite. Since U is infinite, we can pick a $P \in U - \cup\{A_f(Q) \mid f \in W'\}$. In particular, since the nature of (U', W') shows that $U' \subseteq \cup\{A_f(Q) \mid f \in W'\}$, we have $P \notin U'$. As $U \subseteq \cup\{A_f(I) \mid f \in W\}$, there is an $f \in W$ with $P \in A_f(I)$, so that $f(P_P) = f(I_P)$. Now $I \subseteq Q \subseteq P$. Thus $f(P_P) = f(I_P) \leq f(Q_P) \leq f(P_P)$. Equality holds throughout, and so $P \in A_f(Q)$. By choice of P, $f \notin W'$. Considering $(U' \cup \{P\}, W' \cup \{f\})$, and taking the obvious one-to-one correspondence, we have contradicted the maximality of the size of U', and so proved the claim. Let (U', W') be as in the claim. Since $f(Q_Q) \leq$ height Q, and since W' is infinite, there is an integer

$0 \leq n \leq$ height Q, and an infinite subset W'' of W' such that $f(Q_Q) = n$ for all $f \in W''$. Let U'' be the subset of U' corresponding to W'' under the one-to-one correspondence between U' and W'. Then U'' is infinite, and so (Q, U'') is a conforming pair. Let $h = \min W''$. By (d), h is a grade function. For $P \in U''$, if $f \in W''$ corresponds to P, then $P \in A_f(Q)$, so that $h(P_P) \leq f(P_P) = f(Q_P) \leq f(Q_Q) = n$. Since (Q, U'') is a conforming pair, (14.4)(iii) shows $h(Q_Q) \leq n - 1$. But $f(Q_Q) = n$ for all $f \in W''$, so that $h(Q_Q) = n$. This contradiction completes the proof.

So far, we have been considering the set of grade functions on R under the standard partial ordering $\geq$. We can also impose two other partial orderings on this set, which we now define.

Definition. Let f and g be two grade functions on R. By $f \xrightarrow{} g$, we will mean $A_f{}^m \subseteq A_g{}^m$. By $f \Rightarrow g$, we will mean $A_f \subseteq A_g$.

(15.11) Remarks. (15.8) shows that $\xrightarrow{}$, $\Rightarrow$, and $\geq$ are all of the same ilk. Also note that if h is the height function, then $h \xrightarrow{} f$ and $h \Rightarrow f$ for all grade functions f.

(15.12) Proposition. If $f \xrightarrow{} g$ then $f \Rightarrow g$. If $f \Rightarrow g$ then $f \geq g$.

Proof. Clearly $f \Rightarrow g$ implies $f \geq g$. Now suppose that $f \xrightarrow{} g$. For some ideal I and prime $P \in A_f(I)$, we must show that $P \in A_g(I)$. By definition of $A_f(I)$, we have $f(I_P) = f(P_P) = n$ (say). Let $x_1, ..., x_n$ be elements in I whose images in R_P form an f-sequence. Then $n \leq f((x_1, ..., x_n)R_P) \leq$ height $((x_1, ..., x_n)R_P) \leq v((x_1, ..., x_n)R_P) \leq n$. Equality holds throughout, and since $f(P_P) = n$ as well, we see that

$P \in A_f{}^m((x_1, ..., x_n)R)$. Since $f \xrightarrow{} g$, we have $P \in A_g{}^m((x_1, ..., x_n)R)$. However, by (14.18), $A_g{}^m \subseteq A_g$, and so $P \in A_g((x_1, ..., x_n)R)$. Thus

$g(P_P) = g((x_1, ..., x_n)R_P) \le g(I_P) \le g(P_P)$. Thus $g(I_P) = g(P_P)$, showing $P \in A_g(I)$, as desired.

(15.13) Examples. We show that in (15.12), neither converse holds. First, we show that $f \ge g$ does not imply $f \Rightarrow g$. In the example in (15.3) we have $f \ge g$. However $A_f \not\subseteq A_g$ since (15.3) shows that $A_f \cap A_g$ is not a grade scheme. Thus $f \not\Rightarrow g$. We next show that $f \Rightarrow G$ does not imply $f \rightarrow g$. Consider example (15.4). For that f and h, we have that $A_f{}^m \cap A_h{}^m$ is not a grade scheme for f. Thus $A_f{}^m \not\subseteq A_h{}^m$, so that $f \not\rightarrow h$. We will now show that $f \Rightarrow h$. We need $A_f \subseteq A_h$. We need not ·vorry about ideals in any localizations of R other than R itself, because in such localizations, f and h just become the height function. For I any ideal in R, if $P \in A_f(I)$, we need $P \in A_h(I)$. As we may localize at P, the previous sentence shows that we need only worry about that case $P = M$. However, for any I, (14.4)(i) shows that $h(I_M) = h(M_M) = 0$. Thus $M \in A_h(I)$ for all I. This shows that $A_f \subseteq A_h$.

$$\text{If } f \ge g, \text{ then for all } P \in \text{ Spec } R, g(P_P) \le f(P_P) \le \text{height } P.$$

We now show that if we make the stronger assumption that $f \rightarrow g$, then a much stronger conclusion holds (showing that $f \rightarrow g$ is much more stringent than $f \ge g$).

(15.14) Lemma. Suppose f and g are grade functions on R. If $f \rightarrow g$, then for all $P \in$ Spec R, either $f(P_P) = $ height P or $f(P_P) = g(P_P)$.

Proof. Suppose that $f \rightarrow g$, and that $f(P_P) = n < $ height P. Let $x_1, ..., x_n$ be elements in P whose images form a maximal f-sequence in P_P. Then $P \in A_f{}^m(x_1, ..., x_n)$. Since $f \rightarrow g$, we see that $P \in A_g{}^m(x_1, ..., x_n)$. As $n < $ height P, P is not minimal over $(x_1, ..., x_n)$. Therefore, by definition of $A_f{}^m$ and $A_g{}^m$, we must have that $f(P_P)$ and $g(P_P)$ both equal $v((x_1, ..., x_n)R_P)$, and so equal each other.

We can consider questions 1 through 4 with respect to the partial orderings $\rightarrow$ and $\Rightarrow$. We will state some results without proof. As before, W will be a nonempty set of grade functions on R.

(15.15) Remark. We can say the following about upper bounds with respect to $\rightarrow$. Let V be the set of all upper bounds to W under $\rightarrow$ and let h = min V. Then h is the least upper bound to W under $\rightarrow$. Also, $A_h{}^m = \cup A_g{}^m$, $g \in V$. Furthermore, if $\cap A_f{}^m$, $f \in W$, is a grade scheme, then it is a grade scheme for h.

(15.16) Remark. We can say the following about lower bounds with respect to $\rightarrow$. W may not have any lower bounds with respect to $\rightarrow$, even if W is finite. The following are equivalent. (a) W has a lower bound with respect to $\rightarrow$. (b) W has a greatest lower bound with respect to $\rightarrow$. (c) min W is the greatest lower bound of W with respect to $\rightarrow$. Also, for the grade function k, k is the greatest lower bound of W with respect to $\rightarrow$ if and only if $A_k{}^m = \cup A_f{}^m$, $f \in W$. Finally, let h be the least upper bound of W with respect to $\geq$. If W has a lower bound with respect to $\rightarrow$, then h is the least upper bound of W with respect to $\rightarrow$, and $A_h{}^m = \cap A_f{}^m$, $f \in W$.

(15.17) Remark. We can say the following about upper bounds with respect to $\Rightarrow$. W may not have a least upper bound with respect to $\Rightarrow$, and if it does have a least upper bound with respect to $\Rightarrow$, that bound may differ from the least upper bound of W with respect to $\geq$. The preceding sentence is true even when W is finite. Let V be the set of all upper bounds to W with respect to $\Rightarrow$, let h = min V, and let k be the least upper bound to W with respect to $\geq$. Then $h \in V$, and $g \Rightarrow k$ for all $g \in V$. Furthermore, the following are equivalent. (a) W has a least upper bound with respect to $\Rightarrow$. (b) h is the least upper bound of W with respect to $\Rightarrow$. (c) $A_h = \cup A_g$, $g \in V$.

(15.18) We can say the following about lower bounds with respect to $\Rightarrow$. W may not have any lower bound with respect to $\Rightarrow$. If W does have a lower bound with respect to $\Rightarrow$, then for every subset W' of W, min W' is a grade function. Even when W is finite, W may have lower bounds but no greatest lower bound with respect to $\Rightarrow$, and if it does have a greatest lower bound with respect to $\Rightarrow$, that bound may not be min W.

Let U be the set of all lower bounds to W with respect to $\Rightarrow$, and suppose that U is nonempty. Let k = min W and let h be the least upper bound of U with respect to $\geq$. Then $h \in U$ and $k \Rightarrow g$ for all $g \in U$. Furthermore, the following are equivalent. (a) W has a greatest lower bound with respect to $\Rightarrow$. (b) h is the greatest lower bound of W with respect to $\Rightarrow$.

(c) $A_h = \cap A_g$, $g \in U$.

REFERENCES

[B] M. Brodmann, Asymptotic stability of Ass $(M/I^n M)$, Proc. Am. Math. Soc., 74(1979) 16-18.

[BR] M. Brodmann and C. Rotthaus, Local domains with bad sets of formal prime divisors, J. Algebra, 75(1982) 415-419.

[FR] D. Ferrand and M. Raynaud, Fibres formelles d'un anneau local Noetherian, Ann. Sci. Ecole Norm. Sup., 3(1970) 295-311.

[G] A. Grothendieck, Elements de Geometrie Algebrique,IV, seconde partie, Inst. Hautes Etudes Sci. Publ. Math., Presses universitaires de France, Paris, Frince, 1965.

[H] S. Huckaba, On linear equivalence of the P-adic and P-symbolic topologies, J. Pure and Applied Algebra, 46 (1987) 179-185.

[Kp] I. Kaplansky, Commutative Rings, University of Chicago Press, 1974.

[Kz1] D. Katz, Prime divisors, asymptotic R-sequences and unmixed local rings, J. Algebra, 95(1985) 59-71.

[Kz2] D. Katz, Two applications of asymptotic and essential sequences, Houston J. Math, 13 (1987) 65-73.

[KM] D. Katz and S. McAdam, Two asymptotic functions, Comm. in Algebra 17 (1989) 1069-1091.

[KMOR], D. Katz, S. McAdam, J. Okon, and L. J. Ratliff, Jr., Essential prime divisors and projectively equivalent ideals, J. Algebra 109 (1987) 468-478.

[KMR] D. Katz, S. McAdam, and L. J. Ratliff, Jr., Prime divisors and divisorial ideals, J. Pure and Applied Algebra, (in publication).

[KR1] D. Katz and L. J. Ratliff, Jr., U-Essential prime divisors and sequences defined over an ideal, Nagoya J. Math. 103(1986) 39-66.

[KR2] D. Katz and L. J. Ratliff, Jr., On the symbolic Rees ring of a primary ideal, Comm. in Algebra, 14 (1986) 959 - 970.

[M1] S. McAdam, Saturated chains in Noetherian rings, Indiana Univ. Math. J., 23 (1974) 719 - 128.

[M2] S. McAdam, Asymptotic Prime Divisors, Lecture Notes in Mathematics No. 1023, Springer-Verlag, New York, 1983.

[M3] S. McAdam, Grade schemes and grade functions, Trans. Am. Math. Soc., 288(1985) 563-590.

[M4] S. McAdam, Filtrations, Rees rings, and ideal transforms, J. Pure and Applied Algebra, 42(1986) 237-243.

[M5] S. McAdam, Quintasymptotic primes and four results of Schenzel, J. Pure and Applied Algebra, 47 (1987) 283 - 298.

[MR1] S. McAdam and L. J. Ratliff, Jr., Essential sequences, J. Algebra, 95(1985) 217-235.

[MR2] S. McAdam and L. J. Ratliff, Jr., Finite transforms of a Noetherian ring, J. Algebra, 103(1986) 479-489.

[MR3] S. McAdam and L. J. Ratliff, Jr., Persistent primes and projective extensions of ideals, Comm. in Algebra, 16 (1988) 1141 - 1185.

[MR4] S. McAdam and L. J. Ratliff, Jr., Sporadic and irrelevant prime divisors, Trans. Am. Math. Soc., 303 (1987) 311 - 324.

[N1] M. Nagata, On the chain problem of prime ideals, Nagoya Math. J., 10 (1956) 51 -64.

[N2] M. Nagata, Note on a paper of Samuel concerning asymptotic properties of ideals, Memoirs U. Kyoto, series A, 30 (1957) 165-175.

[N3] M. Nagata, Local Rings, Interscience, New York, 1962.

[R1] L. J. Ratliff, Jr., Quasi-unmixed semilocal domains and the altitude formula, American J. Math., 87 (1965) 278 - 284.

[R2] L. J. Ratliff, Jr., On quasi-unmixed local domains, the altitude formula, and the chain condition for prime ideals (II), American J. Math. 92 (1970), 99 - 144.

[R3] L. J. Ratliff, Jr., Essential sequences over an ideal and essential cograde, Math. Z., 188 (1985) 383-395.

[R4] L. J. Ratliff, Jr., On linearly equivalent ideal topologies, J. Pure and Applied Algebra, 41(1986) 67-77.

[R5] L. J. Ratliff, Jr., The topology determined by the symboloic powers of primary ideals, Comm. in Algebra, 13 (1985) 2073 - 2104.

??[R6] L. J. Ratliff, Jr., Δ-Closure operations on ideals and rings, Trans. AMS 312 (1989).

[Re] D. Rees, Rings associated with ideals and analytic spreads, Math. Proc. Camb. Philo. Soc., 89(1981) 423-432.

[S] P. Samuel, Some asymptotic properties of powers of ideals, Annals Math., 56(1952) 11-21.

[Sc] P. Schenzel, Finiteness of relative Rees rings and asymptotic prime divisors, Math. Nachr. 129 (1986) 123-148.

[Sh] K. Shah, Coefficient ideals of the hilbert polynomial and integral closures of parameter ideals, dissertation, Purdue Univ. (1988).

[Sp] R. Y. Sharp, Asymptotic behaviour of certain sets of attached prime ideals, J. London Math. Soc., 34(1986) 212-218.

[V1] J. Verma, On ideals whose Adic and Symbolic topologies are linearly equivalent, J. Pure and Applied Algebra, 47(1987) 205-212.

[V2] J. Verma, On the symbolic topology of an ideal, J. Algebra, 112 (1988) 416-429

INDEX

INDEX TO SYMBOLS